T0309983

Oral Formulation Roadmap from
Early Drug Discovery to Development

Oral Formulation Roadmap from Early Drug Discovery to Development

Edited by Elizabeth Kwong

Registered Offices
John Wiley & Sons, Inc., 111 River Street, Hoboken, NJ 07030, USA

Editorial Office
111 River Street, Hoboken, NJ 07030, USA

For details of our global editorial offices, customer services, and more information about Wiley products visit us at www.wiley.com.

Wiley also publishes its books in a variety of electronic formats and by print-on-demand. Some content that appears in standard print versions of this book may not be available in other formats.

Library of Congress Cataloging-in-Publication Data

Names: Kwong, Elizabeth, 1954– editor.
Title: Oral formulation roadmap from early drug discovery to development / edited by Elizabeth Kwong.
Description: Hoboken, NJ : John Wiley & Sons Inc., 2017. | Includes bibliographical references and index.
Identifiers: LCCN 2016045904 | ISBN 9781118907337 (cloth) | ISBN 9781118907900 (Adobe PDF) | ISBN 9781118907870 (epub)
Subjects: | MESH: Drug Discovery | Chemistry, Pharmaceutical–methods | Clinical Trials as Topic | Dosage Forms | Administratoin, Oral
Classification: LCC RS420 | NLM QV 745 | DDC 615.1/9–dc23
LC record available at https://lccn.loc.gov/2016045904

Cover design by Wiley.
Cover image: © annedde/Gettyimage and Steven Wright/Shutterstock.

Set in 10/12pt Warnock by SPi Global, Pondicherry, India

10 9 8 7 6 5 4 3 2 1

Contents

List of Contributors

Steven Booth Merck Sharpe & Dohme, Hoddesdon, Hertfordshire, UK

Gerard Byrne Merck Sharpe & Dohme, Hoddesdon, Hertfordshire, UK

Lorenzo Capretto Merck Sharpe & Dohme, Hoddesdon, Hertfordshire, UK

Pierre Daublain Discovery Pharmaceutical Sciences, Merck Research Laboratories, Boston, MA, USA

Lee Dowden Merck Sharpe & Dohme, Hoddesdon, Hertfordshire, UK

Kung-I Feng Discovery Pharmaceutical Sciences, Merck Research Laboratories, Kenilworth, NJ, USA

Shayne Cox Gad Gad Consulting Services, Cary, NC, USA

Mengwei Hu Discovery Pharmaceutical Sciences, Merck Research Laboratories, Kenilworth, NJ, USA

Elizabeth Kwong Kwong Eureka Solutions, Montreal, Quebec, Canada

Dennis H. Leung Small Molecule Pharmaceutical Sciences, Genentech, Inc., South San Francisco, CA, USA

Mark McAllister Drug Products Design, Pfizer Ltd., Sandwich, UK

Caroline McGregor Merck Research Laboratories, Kenilworth, NJ, USA

Evan A. Thackaberry Genentech, Inc., Safety Assessment, South San Francisco, CA, USA

Sarah Trenfield Merck Sharpe & Dohme, Hoddesdon, Hertfordshire, UK

Mei Wong Drug Products Design, Pfizer Ltd., Sandwich, UK

Preface

The discovery and development of new drugs is a very complex machine. Despite increasing investments in research and development, the number of new drug approvals has not increased, while the attrition rate of new drug candidates has increased. Many of these challenges are due to failure to properly identify formulations that are translatable from preclinical to the clinic due to lack of effective predictions of therapeutic and toxicological responses in the preclinical stages. Moreover, efforts spent to integrate the formulation scientists in the early discovery that leads to the lead candidate selection had been disappointing. Most of the time, the lack of understanding of the interplay of the physiological system to the formulation contributed to the failure to integrate the right expertise at the right time, which leads to poor clinical successes. The lack of collaboration and proper integration between the formulation and discovery scientists is the root cause of most of the failure in the clinic. Lastly, the understanding of regulatory requirements for formulations also can add to the burden of the timeline and cost of bringing a drug candidate forward.

This book describes and explains key factors that will help determine the types of formulation needed at the different stages of discovery. The considerations of limited amount of API in early stages to the use of the formulation to determine key efficacious or toxicological end point that will not interfere with readouts will be discussed. The formulation selection stage-dependent approach will be detailed up to the planning for the regulatory filing. The interplay of drug metabolism, absorption, and physicochemical properties of the active will be laid out to help understand when a formulation can be improved and when a different lead candidate should be selected. Current formulation approaches based on the biopharmaceutics classification system (BCS) of the lead will be explained. The book will also focus on the relationships between various disciplines like physical chemistry, analytical chemistry, biology, DMPK, toxicology, and medicinal chemistry in determining the appropriate formulation to deliver the candidate in different forms. API sparing approaches including *in vitro* and fit-for-purpose formulation to support first-in-human study will also be covered in the book. Partnership considerations with contract manufacturing organization (CMO) will also be described and shared to increase the probability of meeting tight timelines and to ensure the proper selection of formulation to support an early stage development and how this can impact the late stage development of the drug candidate. Introduction of current formulation approaches including enabling formulations such as solid dispersions used in the industry will widen partnership with emerging innovators and sponsors, making it possible for the otherwise difficult drug candidate to be studied in the clinic.

This book will be the first in detailing the formulation approaches by stage of discovery to early development to help scientists of different disciplines. Practical challenges and solutions will be discussed. The content of the book will guide the proper use of resources to lead scientists to generate the proper database that can help in quick decision-making. The target audience for the book will be drug discovery scientists including medicinal chemists, leaders in pharmaceutical industry (big pharma or start-up companies), and academics who are interested in bringing a potential drug candidate to the clinic. The book will provide real case studies of challenging candidates that allows readers to understand the importance of formulation to their cases. My numerous years (>23 years) working in big pharmaceutical companies, especially the intimate involvement with discovery in the last 15 years of my career and my recent interaction with small- and medium-sized pharmaceutical companies, allowed me to identify collaborators for this book to address the real problems and solutions in drug discovery related to all types of formulations.

The editor wishes to thank all the authors for their expertise in their respective sections and their patience during the revision procedures that were necessary to arrive at this juncture of delivering a well-outlined roadmap.

July 2016   *Elizabeth Kwong*

1

Introduction

Elizabeth Kwong

Kwong Eureka Solutions, Montreal, Quebec, Canada

1.1 Overcoming Challenges in Big Pharma and Evolution of Start-Up Companies

The discovery and development of new drugs is a very complex process. No matter how you implement Lean Six Sigma Black Belt or in-depth data mining into the process, cost and success rate of commercializing drugs had not improved. It was estimated that it takes at least 10 years for a drug to make the journey from discovery to consumer at an average cost of $5 billion (Herper, 2013). Another study conducted by BIO and BioMedTracker (Hay *et al.*, 2011), which collects data on drugs in development, had reviewed more than 4000 drugs from small and large companies that indicated that overall success rate for drugs moving from early stage phase I clinical trials to FDA approval is about 1 in 10, down from 1 in 6 seen in reports earlier. Despite increasing investments in research and development, the number of new drug approvals has not increased, while the attrition rate of new drug candidates has increased.

Recent publication in Fortune entitled "Big Pharma Innovation in Small Places" (Alsever, 2016) quoted several big pharma executives as to the current nature of big pharmaceutical companies where the focus of R&D is diminished to sorting out changes in the company and reprioritizing programs. Furthermore, with investor money flooding in and shift of drug pipelines from internal R&D to start-ups licensing opportunities, big pharma is acquiring

Oral Formulation Roadmap from Early Drug Discovery to Development,
First Edition. Edited by Elizabeth Kwong.
© 2017 John Wiley & Sons, Inc. Published 2017 by John Wiley & Sons, Inc.

small companies at faster pace than before. Small start-ups are now becoming the "new" innovative machines, which offer the high risk–high reward paradigm. According to surveys, last year, 64% of the approved phase I studies originated at a smaller start-ups.

1.2 Overview of Activities Involved in Current Drug Discovery and Development

There had been many surveys that revealed the cause of attrition of molecule in clinical development through the years. The major factors for discontinuation of clinical candidates are lack of efficacy (~30%) and toxicity (~30%). Kola & Landis (2004) further revealed that a 10% drop in attrition in 2000 was partly due to advancement in formulation technologies. Furthermore with increase in molecular obesity in drug candidates in recent years, majority of new drug development is poorly water soluble (Hann, 2011). About 40% of drugs with market approval and nearly 70–90% of molecule in discovery are poorly water soluble, which can lead to low bioavailability with conventional formulations (Kalepu & Nekkanti, 2015). With the introduction of various drug delivery technologies, numerous drugs associated with poor solubility and low bioavailability have been formulated into successful drug products. In fact, recently an increase in NDA file under 505(b)(2) is gaining more importance. New dosage forms with improved solubility and enhanced bioavailability such as prodrugs/active metabolite of drug and reformulation of poorly absorbed drugs using new technologies are turning into lucrative business. According to the Q&A with Ken Phelps, president of Camargo Pharmaceutical Services, which provides services for drug development for 505(b)(2) applications, approximately 20% of new drug approved in 2006 is through 505(b)(2) process. By 2008 more than half of new drug approval was based on 505(b)(2) process (Phelps, 2013).

Poor solubility of development candidates can limit drug concentration at the biological target site, which can lead to loss of therapeutic effect. Increasing the dose can overcome this lack of therapeutic effect but can lead to high variability in absorption, which can be detrimental to the safety and efficacy profile. For these reasons, solubility-enhancement technologies are being used increasingly in the pharmaceutical field. A formulation scientist's approach to solubility enhancement of a poorly water-soluble drug can vary. Often, physicochemical characterization, solid-state modifications, nonconventional formulation technologies, and enabling formulations are often utilized. There are numerous literature resources available to provide guidance toward formulation development from discovery to development of development candidates; however, a single reference where formulation approaches are described in each stage is lacking. This book describes and explains key factors that will help determine the types of formulation needed at the different stages of discovery. The considerations of limited amount of API in early stages to the use of the formulation to determine key efficacious or toxicological end point that will not interfere with readouts will be discussed. The formulation selection stage-dependent approach will be detailed up to the planning for the regulatory filing. The interplay of drug metabolism, absorption, and physicochemical properties of the active will be laid out to help understand when a formulation can be improved and when a different lead candidate should be selected. Current formulation approaches based on the biopharmaceutics classification system (BCS) of the lead will be explained. The book will also focus on the relationships between various disciplines like physical chemistry, analytical chemistry, biology, DMPK, toxicology, and medicinal chemistry in determining the appropriate formulation to deliver the candidate in different forms. API sparing approaches including fit for purpose formulation to get candidates into development will also be covered in the book. Each stage of formulation (see Table 1.1) development has its goals, degree of complexity, and increasing availability of information, which ultimately leads to candidate that will have properties that can be administered in humans.

1.3 Value of the Right Formulation at the Right Time

Many of the discovery challenges are due to failure to properly identify formulations that are translatable from preclinical to clinical due to lack of effective predictions of therapeutic and toxicological responses in the preclinical stages. Moreover, efforts spent to integrate the formulation scientists in the early discovery that leads to the lead candidate selection had been disappointing. Most of the time, the lack of understanding of the interplay of the physiological system and physicochemical properties of the molecule to the drug delivery system contributed to the failure to integrate the right expertise at the right time, which leads to poor clinical successes. The lack of collaboration and proper integration between the formulation and discovery scientists is the root cause of most of the failure in the clinic. Lastly, the understanding of regulatory requirements for formulations also can add to the burden of the timeline and cost of bringing a drug candidate forward.

Although discovery starts off with the structure-based drug design, a better design of drug should be an understanding of how the biological effect is influenced by physicochemical properties, PK of the drug, and pharmaceutical delivery system. Optimization of the API via salt formation or co-crystal and physical changes such as particle size reduction through milling or formation of amorphous dispersions are often employed to improve oral bioavailability of insoluble compounds. These approaches can be applied even at the lead

Table 1.1 Activity definition from discovery to preclinical development.

Early discovery (lead ID/target validation)	Lead optimization/candidate nomination	Preclinical development to phase I

• Un-optimized phase of the molecules
• Limited compound supplies
• HTS-short timeline and high number of leads being screened
• Pharmacology studies (target engagement; efficacy studies)
• *In silico* tox screen
• *In vitro* metabolism
	• More API available
	• Chronic efficacy/biomarker studies
	• Initiate physicochemical characterization
	• Assess developability of the candidate
	  • Synthetic scale-up (~1–10 g)
	  • Potential dose
	  • Dose range finding (DRF) studies
	  • ADME
		• GLP tox study
		• Polymorph/salt screen
		• Scale-up of API

• Standardized solutions for *in vitro* HTS and *in vivo* PK screen
• No vehicle screen
• Usually contains DMSO or other standardized cosolvent vehicle (such as PEG/EtOH), low dose PK with IV/oral for %F
	• Dose range finding to identify exposure multiples
	• Resort to vehicle screen decision tree[a,b,c,d]
	• Goal of formulation selection are:
	  o Vehicles do not have any biological adverse effect
	  o Achieve exposure at the highest toxicological dose
	  o Can reach up to 2 g/kg
	  o Key is to identify adverse effects
		• Vehicle identified and dose range identified for GLP tox
		  o Repeat preparation of vehicle using optimized API
		  o Characterize physical properties of API in vehicle
		  o Meet GLP requirements

Pharmacology studies— needed a sustained plasma level use of Alzet Osmotic pumps[e]	PK–PD studies—use solution at low dose and suspension at high dose to assess relationship	CTM development—based on physical properties, such as flow, stability, particle size, and BCS, bioavailability

[a] Higgins *et al.* (2012). [b] Maas *et al.* (2007). [c] Li & Zhao (2007). [d] Palucki *et al.* (2010).
[e] Neervannan (2006).

identification if a candidate is deemed to show some potential. Various available formulations are discussed for early discovery in Chapter 2. This chapter will explain which formulation will be suitable at what stage and what features of the drug might suggest one technology over another. Chapters 3 and 4 deal with the different toxicology studies in relations to what formulation will be suitable. Following the development of suitable formulation to deliver required exposure in the early stage of discovery, this will then provide adequate safety assessment and risk of the candidate before proceeding to the more expensive clinical trials. Following this stage, Chapter 7 will cover the formulation technologies that will be scalable to support the first clinical trial study.

Selecting a suitable formulation for your drug candidate can be complicated. Publications on formulation options for poorly soluble drugs are widespread. Each publication would have its approaches with decision trees and had shown proof of success that suits the specific pharmaceutical support system. In other words, taking this approach to another company with a different support function may not work. In my years of experience, to properly select the "right" formulation for a specific compound will still need input from a formulation scientist. This will be someone who poses the breadth of knowledge that can span from understanding of the physiological environment, pharmacology, and physicochemical properties of the molecule that will be intended for development. First to note here is the dose that will be required to be formulated, since solubilization techniques will have their limitation if the doses needed will be high. For example, at the lead optimization stage where safety of the candidate will need to be assessed, high doses are usually expected, and no means of solubilization can be possible that uses excipients that are inert unless your candidate is truly water soluble, which is very rare. It is also worth noting that the term "solubilization" is for the candidate to be soluble in the vehicle or mixture, and this does not include the fact that once this formulation is dosed, solubilization in the physiological environment may pose another hurdle that still can limit the absorption of the drug. This then leads to the question of what is the solubility of this molecule in the physiological milieu? One has to consider the micro-environment that may not be visible and static as we would envision during an *in vitro* test. For example, size reduction technology, which is also one of the solubilization techniques, is used to improve bioavailability. This technology is easy to achieve but may not be applicable to a large proportion of poorly soluble compounds. Evaluation of agglomeration potential of the molecule, understanding of the interplay of the excipients with the physical environment, and stability of the particle, molecule, and crystalline form are required. Another tool is the use of lipid technologies, which uses lipids as primary ingredient to deliver the water insoluble molecule. Lipid formulations are more complex and can produce micelles and microemulsions and will need a formulator to understand how each component of the mixture can ensure the target performance of the molecule from the *in vitro* to the *in vivo* environment. Most of the ingredient may be limited by the amount that can be administered in a preclinical study. At the same time, getting the number of additives together can result in a very viscous vehicle that may itself produce some challenge in a multiple day dosing during a toxicology study. Furthermore, use of such formulation for clinical supplies poses other challenges including use of soft gelatin capsule that can be in an appropriate size for dosing in patients and can be costly.

An important strategy to consider for your formulation selection is simplicity. Try to understand the criticality of solubilization to permeability/metabolism.

In some cases where the molecule is poorly soluble, the oral absorption is still acceptable when given a suspension where the only solubilization was the use of a low concentration of surfactant as a wetting agent aid. This approach can provide a PK profile that will have less C_{max} to C_{trough} ratio and can mitigate some of the adverse effects related to high plasma levels. At the same time this may provide sustain release if solubilization of the molecule is slow and the absorption window is wide. To manage the reproducibility of the PK profile, it will be important to properly characterize the suspension including the form and particle size of the compound in suspension. Such formulation approach in preclinical can also translate into a simple blend in a capsule that can be used in clinical formulation. On the other hand, if the molecule is being metabolized or transported at specific dose or species, formulation may not provide the solution even with the help of permeability enhancers. This is part of the reason why optimal drug-like properties are significant in drug discovery to minimize the complexity of downstream activities.

This book will be the first in detailing the formulation approaches by stage of discovery to early development to help scientists of different disciplines. Practical challenges and solutions will be discussed. The content of the book will guide the proper use of resources to lead scientists to generate the proper database that can help in quick decision making. The target audience for the book will be drug discovery scientists including medicinal chemists, leaders in pharmaceutical industry (big pharma or start-up companies), and academics who are interested in bringing a potential drug candidate to the clinic.

Partnership considerations with contract manufacturing organization (CMO) will also be described and shared to increase the probability of meeting tight timelines and to ensure the proper selection of formulation to support an early stage development and how this can impact the late stage development of the drug candidate. Introduction of current formulation approaches including enabling formulations such as solid dispersions used in the industry will widen partnership with emerging innovators and sponsors, making it possible for the otherwise difficult drug candidate to be studied in the clinic.

References

J. Alsever, "Big Pharma Innovation in Small Places," http://Fortune.com, May 13, 2016 (accessed October 3, 2016).

M.M. Hann, Molecular obesity, potency and other addictions in drug discovery. *Med. Chem. Commun.* 2011, **2**, 349–355.

M. Hay, J. R., D. Thomas, J. Craighead, "BioMedTracker Clinical Trial Success Rate Study," *Presented at the BIO CEO & Investor Conference*, New York City at the Waldorf Astoria, February 15, 2011.

M. Herper, "The Cost of Creating a New Drug Now $5 Billion, Pushing Big Pharma to Change," http://Forbes.com, August 11, 2013 (accessed October 3, 2016).

J. Higgins, M.E. Cartwright, A.C. Templeton. Foundation review: progressing preclinical drug candidates: strategies on preclinical safety studies and the quest for adequate exposure. *Drug Discov. Today.* 2012, **17**(15–16), 828–836.

S. Kalepu, V. Nekkanti, Insoluble drug delivery strategies: review of recent advances and business prospects. *Acta Pharm. Sin. B.* 2015, **5**(5), 442–453.

I. Kola, J. Landis, Can the pharmaceutical industry reduce attrition rates? *Nat. Rev. Drug Discov.* 2004, **3**, 711–715.

P. Li, L. Zhao. Developing early formulations: practice and perspective. *Int. J. Pharm.* 2007, **341**, 1–19.

J. Maas, W. Kamm, G. Hauck. An integrated early formulation strategy—from hit evaluation to preclinical candidate profiling. *Eur. J. Pharm. Biopharm.* 2007, **66**, 1–10.

S. Neervannan Preclinical formulations for discovery and toxicology: physicochemical challenges. *Exp. Opin. Drug Metab.* 2006, **2**(5), 715–731.

M. Palucki, J.D. Higgins, E. Kwong, A.C. Templeton. Strategies at the interface of drug discovery and development: early optimization of the solid state phase and preclinical toxicology formulation for potential drug candidates. *J. Med. Chem.* 2010, **53**, 5897–5905.

K. Phelps, "2012 505(b)(2) Approval—Record Year. 505(b)(2) Development and Services," January 2, 2013.

2

Lead Identification/Optimization

Mei Wong and Mark McAllister

Drug Products Design, Pfizer Ltd., Sandwich, UK

2.1   Introduction

Over the last two decades, the introduction of high-throughput screening (HTS) and combinatorial chemistry has changed the drug discovery process by enabling the rapid evaluation of large number of compounds against targets of interest (Bajorath, 2002; Hefti, 2008; Hughes *et al.*, 2011). In the past, selection of compounds for progression (lead identification) focused mainly on affinity and selectivity. However, it has since been recognized that the physicochemical properties (such as solubility and lipophilicity) of a compound play a significant role in whether the compound progresses to be a successful drug candidate. To ensure that leads selected for progression have the right absorption, distribution, metabolism, and excretion (ADME) properties, rule-based systems such as the "rule of five" have been used to predict the drug-likeness of a

Oral Formulation Roadmap from Early Drug Discovery to Development,
First Edition. Edited by Elizabeth Kwong.
© 2017 John Wiley & Sons, Inc. Published 2017 by John Wiley & Sons, Inc.

compound and guide the selection of compounds for progression (Lipinski *et al.*, 1997).The "rule of five" was developed based on a review of compounds that have successfully progressed into clinical studies and stipulates that for an orally active compound to be successful; it should not violate more than one of the following criteria:

- No more than five hydrogen bond donors (the total number of nitrogen–hydrogen and oxygen–hydrogen bonds)
- No more than 10 hydrogen bond acceptors (all nitrogen or oxygen atoms)
- Molecular weight of less than 500
- Octanol–water partition coefficient ($\log P$) not greater than five

Despite the implementation of "rule of five" type filters to lead selection, a relatively high proportion of drugs entering clinical studies fail to reach the market (Hann, 2011). As a result, alternative methods such as "quantitative estimate of drug-likeness" (QED) have been introduced (Bickerton *et al.*, 2012). QED measured drug-likeness based on the concept of desirability and enabled values for multiple molecular properties to be combined into a single measure of compound quality using a desirability function.

Once the leads are selected, the optimization process starts whereby the weaknesses of the compound are improved, while maintaining the favorable properties of the compound (Hughes *et al.*, 2011) such that the compound entering clinical studies has a good balance of *in vitro* properties and ADME properties.

2.2 Early Characterization of Compounds

During this stage, preformulation data generated on the leads are used to identify developability risks and guide molecular structure modifications. The key challenge for the formulation scientist at this stage is the limited information available on the compound and limited bulk (if any) available of experimentation. Therefore, during the early stages of lead identification and lead optimization, computational modeling and HTS play a crucial role in assessing the physicochemical and pharmaceutical properties of the compound. As the compound progresses through to the later stages of lead optimization and larger quantities of material are available, focused experimentation can be used to answer specific questions about the compound as well as improve the quality of the data generated.

2.2.1 Preformulation

High oral bioavailability is often an important goal for drug development. Therefore, it is important to gain sufficient understanding of the properties that can limit oral bioavailability. In order to obtain an accurate assessment of the biopharmaceutical properties of a compound, the key physicochemical parameters determined during the preformulation stage are solubility, lipophilicity, pK_a, and permeability.

2.2.1.1 Solubility

For orally absorbed drugs, the compound must be dissolved and in solution for absorption to occur. With the increasing number of compounds with poor solubility, solubility-limited absorption has become one of the main reasons for poor bioavailability in the clinic (Di *et al.*, 2012). Solubility-limited absorption is even more of a problem during the preclinical stage for assessment of safety issues, especially where high doses are required. As a result, the importance of conducting solubility studies during the drug discovery stage is well recognized.

During the early stages of lead identification where solubility experimentation is not feasible due to the large numbers of compounds being screened and lack of material, computational models may be used for solubility prediction. These computational models range from simple models using semiempirical equations based on physicochemical properties such as $\log P$ and pK_a to more complex models based on molecular properties such as molecular weight, polar surface area (PSA), and hydrogen-bonding capacity. Although solubility predictions are useful, the accuracy of these models is variable and highly dependent on the training sets used.

Once material is available, experimentation can be conducted to more accurately determine the solubility of a compound. A range of different solubility assays can be conducted depending on the development stage of the compound, amount of material available for experimentation, and the purpose of the data generated. For example, high-throughput (HT) kinetic solubility screens can be used to help compound selection, while equilibrium solubility experiments can be used to help biopharmaceutical predictions and guide formulation development.

2.2.1.1.1 *Kinetic Solubility*

Kinetic solubility assays are typically conducted during lead identification as the assays use stock solutions (e.g., DMSO stock solution), which are readily available at this stage. In addition, the assay uses minimal material, and the format of the assay readily lends itself to automation and integration into the HTS process (Lipinski *et al.*, 2001).

A typical kinetic solubility study would involve the addition of small volumes of stock solution to media to form a supersaturated solution. The solution is then incubated for a short period of time to allow precipitation of the compound. The amount of compound remaining in solution is then analyzed by UV or nephelometry (Avdeef and Testa, 2002; Kerns *et al.*, 2008).

While kinetic solubility is an indicator as to whether a compound may have solubility issues, studies have shown that kinetic solubility tends to overestimate

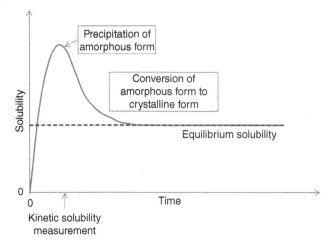

Figure 2.1 Precipitation of amorphous form resulting in over prediction of solubility.

the actual solubility of a compound (Saal and Petereit, 2012; Sugano *et al.*, 2006). This difference could be a result of several factors including the short incubation time and precipitation of an amorphous or metastable solid form (Figure 2.1).

2.2.1.1.2 Equilibrium/Thermodynamic Solubility

Equilibrium or thermodynamic solubility of a compound is defined as the maximum concentration of a compound, which, at a defined temperature and pressure in a given solvent, is thermodynamically valid as long the solid phase exists in equilibrium with the solution phase (Murdande *et al.*, 2011). While equilibrium solubility is considered the "gold standard" for determining the solubility of a compound, it is less commonly used in early lead identification due to the higher bulk and resource requirements as well as longer turnaround times.

Equilibrium solubility measurements are conducted by adding excess solid material to the buffer and shaking the resulting suspension for a predetermined temperature for a defined time (typically between 24 and 48 h). The remaining solid at the end of the experiment is removed, and the amount in solution is analyzed to obtain the equilibrium solubility value of the compound. As the solid form (crystallinity and polymorphic form) of the material may change during the experiment, characterization of the remaining solid is important when reviewing solubility data.

In addition to aqueous buffers, equilibrium solubility studies are frequently conducted using simulated gastric and intestinal fluid (SGF, FaSSIF, and FeSSIF). Solubility results from these studies are used as inputs into *in silico* models such as GastroPlus™ and Simcyp* to help predict *in vivo* performance of the compound.

2.2.1.1.3 Pseudo-Kinetic Solubility

To bridge the gap between the kinetic and equilibrium solubility assays, scientists at Pfizer developed the "pseudo-kinetic solubility" screen. Like the kinetic screen, the pseudo-kinetic screen starts with pre-dissolved compound and can be easily automated. However, the pseudo-kinetic screen has a longer incubation time (20 h), and the screen plate was modified to enable information on the solid form to be obtained using polarized light microscopy (PLM) (Sugano *et al.*, 2006).

Figure 2.2 compares the correlation between the solubility values obtained from the kinetic screen and pseudo-kinetic screen against values obtained via equilibrium solubility studies.

2.2.1.2 pK_a

The acid–base dissociation constant (pK_a) of a compound is used to understand the ionization state of a compound in solution at a particular pH. The pK_a is important as it can influence the solubility, lipophilicity, and permeability of a molecule, especially when ionized at physiologically relevant pH of 2–8 (Manallack, 2007). A molecule in its charged state will show higher solubility than in its uncharged state but, conversely, will have lower permeability. In other words, the permeability and/or solubility of a compound can be altered by the introduction or modification of ionizable groups.

In early lead identification, software packages such as ACD/Labs can be used to predict pK_a values of a compound. pK_a values can also be determined experimentally using either titration (potentiometric or UV spectral detection) or capillary electrophoresis (CE) (Wan *et al.*, 2003).

2.2.1.3 Lipophilicity

The lipophilicity of a compound represents the affinity of the compound for oils, fats, and nonpolar solvents and is determined by either partition coefficient (log$P$) or distribution coefficient (log$D$) measurements using two non-miscible solvents, typically *n*-octanol and water. Log$P$ values are the ratio of concentrations of unionized drug in the octanol–water system, while log$D$ values are the ratio of concentrations of both unionized and ionized drug in the octanol–water system and are affected by the pH of the system. Log$P$/D measurements are usually conducted by adding dissolved compound into a flask containing both octanol and water and shaking the flask until equilibrium is achieved. The concentration of compound in each solvent is then quantified using an appropriate technique such as UV.

The marketed 96-well format octanol–water shake flask method by Analiza, Inc (www.analiza.com) provides log$D$ ranges of –3 to 4. Another common approach is the use of HPLC retention times in relation to a set of standards with known log$D$ to predict an approximate log$D$ for the compound of interest (Yamagami *et al.*, 2002).

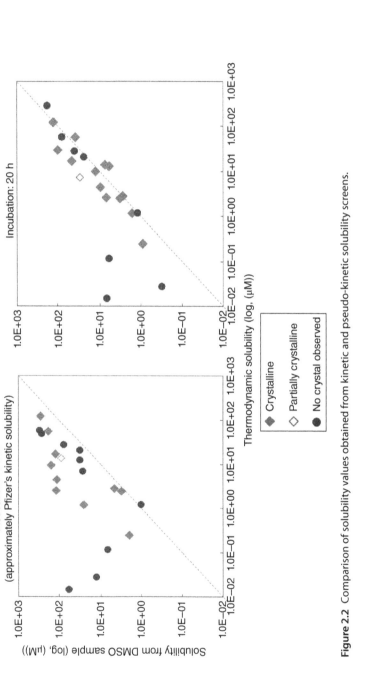

Figure 2.2 Comparison of solubility values obtained from kinetic and pseudo-kinetic solubility screens.

Furthermore, the lipophilicity of the compound influences the permeability of the compound and transport across the gastrointestinal tract (GIT) and blood–brain barrier (see Section 2.2.1.4). In addition, it can also be used during formulation development to help with selection of solubilizing formulations (e.g., use of self-emulsifying drug delivery systems for compound with high log$P$/D).

2.2.1.4 Permeability

Following oral administration, compounds dissolved in gastrointestinal (GI) fluids have the potential to be absorbed via a variety of mechanisms that involves either passive diffusion or active transport (Figure 2.3). For the majority of drug compounds, the main route of absorption occurs via passive diffusion through the transcellular pathway. The transcellular diffusion rate is mainly determined by the rate of transport across the apical cell membrane and is controlled by the lipophilicity and ionization state of the compound.

This interrelationship of pK_a, log$D$/P, and pH of the absorption site in the GIT forms the basis of the pH-partition theory (Shore *et al.*, 1957). The theory states that transcellular diffusion of a drug molecule through the lipid bilayer of the intestinal membrane can only occur if the molecule is in its unionized state. Therefore, absorption of weakly basic drugs is favored in the small intestine where the pH is higher, and therefore, a larger proportion of unionized drug will be available for absorption. Conversely, absorption of weak acids will be favored in the stomach where pH is lower. However, in reality, the small

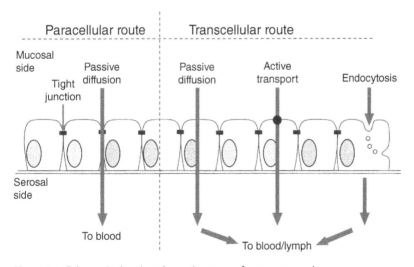

Figure 2.3 Schematic showing absorption routes for a compound.

intestine continues to be the main site of absorption for all drugs due to the larger surface area available for absorption compared with the stomach and colon (Hurst *et al.*, 2007).

During early drug discovery, permeability estimations can be made using the PSA of a compound. Results from work by Palm *et al.* (1996, 1997) have shown that compounds with PSA < 60 Å^2 show complete absorption, while compounds with PSA > 140 Å^2 have unacceptably low absorption (Artursson and Bergström, 2004). More recently, studies looking at the use of surface activity profiling (relationship between surface activity and surface pressure of a drug solution) and surface tension on drug permeability were undertaken as part of the innovative tools for oral biopharmaceutics (OrBiTo) project (Bergström *et al.*, 2014).

In vitro permeation studies can be conducted using cell monolayers (e.g., Caco-2 or MDCK) or artificial membranes (e.g., parallel artificial membrane permeability assay (PAMPA)). Figure 2.4 shows the typical setup for the permeability assay. The dissolved compound is added to the apical chamber at the start of the experiment, and the apparent permeability (P_{app}) is determined using Equation 2.1:

$$P_{app} = \frac{dQ}{dtAC_0} \tag{2.1}$$

where dQ/dt = permeability rate; A = surface area of membrane; and C_0 = initial concentration in the apical chamber.

Caco-2 cell monolayers are most commonly used for permeability screening due to their morphological and functional similarity to the human intestine (Varma *et al.*, 2004). The key disadvantage of Caco-2 permeability is the intra- and interlab variability of P_{app} values due to variability in experimental conditions (e.g., pH and/or ionic conditions of the media, shaking rate) or cell culture conditions (Yamashita *et al.*, 2000). Therefore, it is important that the P_{app} values must be used alongside reference values from internal standards.

Most Caco-2 permeability experiments are conducted using either pH 6.5 or 7.4 in the apical chamber. As the degree of ionization can have a significant impact on permeability, it is important to consider the pK_a of the compound

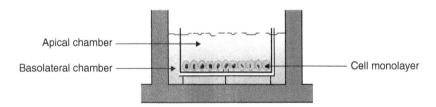

Apical chamber

Basolateral chamber

Cell monolayer

Figure 2.4 Typical setup for permeability assay.

alongside P_{app} values. For example, metoprolol has a P_{app} of 7.8×10^{-6} cm/s at pH 6.5, while at pH 7.4, the P_{app} increases to 39.7×10^{-6} cm/s.

2.3 Formulation Approaches in Drug Discovery

Formulation development during the early stages of drug discovery is often complicated by the limited availability of drug, which contributes to an incomplete physicochemical and biopharmaceutical profile. Solubility and permeability properties are often unfavorable and necessitate the use of approaches over and beyond simple solution or suspension formulations to achieve the required levels of exposure. Numerous literature reports (Chaubal, 2004; Li and Zhao, 2007; Maas *et al.*, 2007; Niwa and Hashimoto, 2008; Saxena *et al.*, 2009; Wilson, 2010) have described significant formulation efforts during the candidate-profiling phase, but only a few publications describe comprehensive formulation strategies to support early discovery *in vivo* studies (Kwong *et al.*, 2011). These studies can be classified into three main types: (i) pharmacokinetic (PK) study, (ii) pharmacodynamic (PD) study, and (iii) toxicokinetic/dose range finding studies to support toxicological studies (Shah and Agnihotri, 2011). These studies are carried out sequentially during the lead identification to lead optimization stages. PK studies are an essential first step to evaluate the fate of the drug after administration and to provide information on the ADME properties of a lead molecule or series. By contrast, PD studies evaluate the effect of drug on the body and are used to elucidate concentration drug activity relationships and to establish any correlation of PK to biological activity or efficacy. Toxicokinetic or dose range finding studies are performed to determine exposure/dose relationships and to develop an understanding of the maximal absorbable dose. These studies are followed by multiple dosing studies, which are used to establish the no observable effect level (NOEL) or the no observable adverse effect level (NOAEL) for a new chemical entity. The formulation strategies for toxicology studies will be discussed in a subsequent chapter of this book.

2.3.1 PK/PD Studies of Lead Compounds: Formulating for Preclinical Development

Throughout the early discovery phase, formulation expertise is required to enable PK and early biology studies in preclinical species to establish critical ADME parameters and proof of pharmacology for a lead series. As a molecule progresses through lead optimization into exploratory toxicology and then regulatory safety assessment studies, the need to develop effective formulations to deliver sufficient exposure is critical to the success of the program. The requirement to drive exposure in toxicology to levels that will provide adequate safety coverage for future clinical studies is often a considerable development hurdle to overcome. This spectrum of preclinical formulation activities is made even more difficult by the challenging physicochemical properties of typical lead candidate molecules. It is clear from experience across the industry that the exploration of novel biological target space and pharmacological mechanisms combined with the use of high-throughput modern drug discovery approaches to yield highly potent and selective molecules has provided chemical substrate, which has become increasingly difficult to formulate due to poor molecular properties such as hydrophobicity and low aqueous solubility. The additional complexity associated with such substrate has forced the industry to apply formulation resource much earlier in the discovery process. Past practices such as "disperse and dose" (Maas *et al.*, 2007) where discovery biologists, chemists, or drug metabolism scientists would formulate simple solutions and suspensions for screening PK or early PD studies have been largely superseded by the integration of candidate-enabling formulation teams into the multidisciplinary discovery teams. This is evident from the number of publications from groups in major pharma companies with specific roles for discovery support, for example, Developability Assessment group at Novartis (Saxena *et al.*, 2009), Research Formulation at Pfizer, Discovery Pharmaceutics at BMS (Chen *et al.*, 2012), and Basic Pharmaceutical Sciences group at Merck (Palucki *et al.*, 2010).

Some common challenges are encountered during preclinical formulation development for both PK and PD studies. Quantities of API are often limited (10–20 mg is not typical for formulation development), and quality of drug substance can be variable in terms of purity, particle size, and solid form attributes. Timelines are also compressed with discovery teams requiring formulations with a timeframe of days rather than weeks and certainly not months' worth of notice. This of course can be tackled by efficient formulation development using miniaturized formulation technologies and standardized formulation protocols, but it does require the availability of physicochemical characterization data and a fit-for-purpose analytical method. There are also some key differences between PK and PD studies with regard to design and target parameters, which place different constraints on the formulation approach, which can be adopted. Generally, PD studies are more complex to deal with due to their longer duration, multiple dosing, and the need to avoid any biological activity associated with formulation excipients. Table 2.1 summarizes these and a number of additional differentiating aspects between PK and PD studies (Maas *et al.*, 2007).

In vitro metabolism data (obtained from liver microsomes or hepatocytes from mouse, rat, dog, or human) (Asha and Vidyavathi, 2010; Naritomi *et al.*, 2003; Obach *et al.*, 1997; Parkinson *et al.*, 2010) are a key component of the lead selection and optimization process. Compounds emerging from metabolic screens with limited predicted PK liability or those for which data

Table 2.1 Differences between PK and PD studies.

PK studies	PD studies
Cassette dosing feasible	Cassette dosing not appropriate
Single administration	Often multiple or chronic administration required
PD effect of solvent can be acceptable	PD effects of solvents not acceptable
Normal, healthy animals used	Specialized or disease animal models used
Evaluation rapid	Evaluation complex and longer
No vehicle control	Vehicle control groups required
No requirement for PD evaluation	Parallel PK analysis often conducted

Source: Adapted from Maas *et al.* (2007).

obtained from screening systems such as UGTs have a low confidence in prediction (Ritter, 2007; Soars *et al.*, 2002) are selected for PK assessment in whole animal studies. Rats are commonly used as the species of choice at this stage due to ease of dosing and sampling, although mice can be used if later efficacy models are already established in this species or if the rat shows species-specific metabolism (Saxena *et al.*, 2009). The objective of these studies is to provide data on intrinsic PK parameters (plasma half-life, clearance, and bioavailability) and fully profile the ADME properties of candidate molecules. Data from such early PK studies are used to inform the design of follow-on pharmacological and toxicology studies.

Typically, early PK studies will be designed with IV and oral legs at relatively low doses (0.5–10 mg/kg in rats) and will use simple solution or suspension formulations where possible. The flowchart in Figure 2.5 shows the principles used to guide the design of IV formulations for preclinical PK studies in the authors' labs.

The feasibility of simple solubilization approaches based on pH adjustment can be readily assessed from basic physicochemical parameters such as thermodynamic solubility and calculated or measured pK_a. For nonionizable compounds, cosolvents are often employed as they can enhance the solubility of nonpolar molecules by several orders of magnitude (Lee *et al.*, 2003). In a study of 300 compounds from the Pfizer discovery pipeline formulated in 2000, approximately 38% were found to have solubilities lower than 30 μg/mL, and overall approximately 33% were unionized. The authors of this study proposed a complex formulation decision tree, which accounted for the ionization state of the molecule and solubility in mixed aqueous–cosolvent systems containing one or more cosolvents selected from dimethylacetamide, ethanol, propylene glycol, and PEG 400. Using this empirical approach the authors were able to

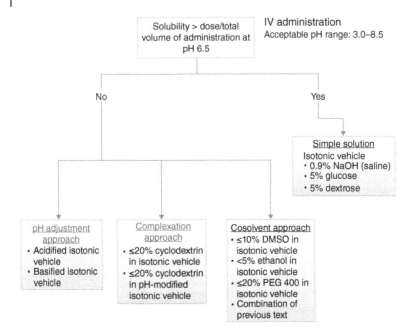

Figure 2.5 Flowchart to guide IV formulation for preclinical *in vivo* studies.

solubilize over 80% of submitted compounds in a vehicle, which contained cosolvents in acceptable quantities and with at least a 20% aqueous component. Such an empirical scheme can be adapted to suit a high-throughput-miniaturized screen and offers a simple route for cosolvent selection. An alternative to the empirical approach is to utilize computational tools to predict cosolvent solubility. One such example of this approach is the use of the quantum chemistry program, conductor screening model for real solvents (COSMO-RS) tool by Pozarska *et al.* (2013). COSMO-RS calculations are independent of any molecular and structural information, thus making it suitable for use in early discovery where as stated previously, compound physicochemical characterization information can be limited. COSMO-RS was evaluated for the potential to accelerate the selection of excipients for preclinical formulation development. The excipient solubility predictions obtained from COSMO-RS for seven compounds with low aqueous solubility (<1 mg/mL) and solubility-limited exposure in preclinical studies were compared with experimentally obtained solubilities for sixteen cosolvents. Overall, five of the seven compounds tested showed good agreement between the COSMO-RS predicted solubilities and experiment solubilities. COSMO-RS correctly ranked the majority of the sixteen excipients. Therefore, when developing a formulation, excipients with predicted low solubility can be quickly eliminated from the excipient selection process, thus saving API, time, and experimental resource. A major limitation of the cosolvent approach for IV delivery is potential drug precipitation on administration. This may be more of a concern for IV bolus administration than infusion due to the shorter time but can result in local adverse reactions and variable plasma levels (Li and Zhao, 2007). Understanding the supersaturation phase of the drug of interest in the cosolvent vehicle and propensity to precipitate on dilution with plasma is an important consideration during development of IV cosolvent formulations.

For molecules not suitable for pH or cosolvent solubilization approaches, complexation using cyclodextrins can be considered. While requiring more compound to support formulation development, modified β-cyclodextrin derivatives such as hydroxypropyl-β-cyclodextrin (HP-β-CD) and SBEβCD have good solubility and safety profiles, and multiple publications are available, supporting their use for preclinical formulation (Koltun *et al.*, 2010; Li and Zhao, 2007; Shah *et al.*, 2014). It is important to develop an understanding of binding constants with cyclodextrins so as to avoid any impact of the cyclodextrin on the distribution of the molecule of interest in plasma (Buggins *et al.*, 2007). Surfactants can also be used to assist solubilization for preclinical IV formulations, but historically their use has been limited due to concerns relating to allergic reactions in some animal species. Progress with surfactant research has provided new materials for the preclinical formulation scientist to consider with functionalized polymeric materials such as Solutol HS 15 (macrogol 15 hydroxystearate), vitamin E TPGS (D-alpha-tocopheryl polyethylene glycol 1000 succinate), and pluronics/poloxamers (block copolymers of polyethylene glycol and polypropylene glycol) offering alternatives to older surfactants such as polysorbate 80 or the cremophor series. More sophisticated formulation approaches using liposomal or nanoparticle systems are not typically used in early PK studies due to the impact of the delivery system on biodistribution and the complexity associated with the development of these systems.

It should be noted that parenteral delivery is not restricted to just early PK studies. In early PD biology studies, the delivery of lead molecules is often not optimized, and parenteral delivery, via either intraperitoneal subcutaneous or intravenous routes, provides a practical option for probing PK/PD relationships. While these routes have significant limitations in terms of vehicle selection and acceptability, they do open up the possibility of continuous delivery through mini-pumps or sustained infusions. The Alzet* osmotic pump is one such example that can be implanted subcutaneously (or if required in the intraperitoneal cavity) and can deliver extended infusions over a 7–14-day period if required. Volume limitations of the pumps (around 0.2 mL in mice and 2 mL in rats) are significant constraints when considering their application in rodent species, and it is necessary to adapt the delivery vehicle to solubilize high concentrations of the active of interest (Gullapalli *et al.*, 2012; Neervannan, 2006). Osmotic pumps have been shown to deliver more complex formulations with nanosuspensions of fenofibrate successfully delivered from subcutaneously implanted devices in mice (Hill *et al.*, 2013).

The oral leg of early PK studies is intended to provide an indication of bioavailability and the extent of any absorption limitation relating to dissolution and/or solubility of the lead series. As already discussed, the preponderance of poor solubility candidates in discovery pipelines drives the use of enabled or bio-enhanced formulation even at this early stage of preclinical development. Solution formulations can be used, but as with their IV counterparts, care is needed to ensure that precipitation on mixing with GI luminal fluids does not occur. A very commonly used approach is to dose a simple polymer or surfactant-stabilized suspension. In this case, the polymers used (such as methyl- or hydroxypropyl cellulose) aid suspension and homogenization, while the surfactants aid dispersion and wettability. Physical stability of simple suspensions needs to be confirmed during early formulation development, and monitoring solid form change over this period is also important. Particle size control for suspension preparation can range from simple techniques such as mortar and pestle grinding through to the use of milled/micronized material or using processing method such as high-pressure homogenization or wet milling to achieve consistent particle attrition.

If these simple formulation approaches are insufficient to deliver consistent, reproducible exposure *in vivo*, bio-enhanced formulations such as nanoparticles, solubilizing systems (cyclodextrins, self-emulsifying drug delivery systems, etc.), or stabilized amorphous systems can be considered. At this stage of pre-clinical development, manipulation of solid form (apart from *in situ* salt formation (Tong and Whitesell, 1999)) is not typically used due to time constraints and the limited availability of API. Similar restrictions apply to the development of prodrugs, specifically for the purpose of overcoming exposure limitations at this very early stage. The selection of an appropriate bio-enhanced approach is usually driven by an analysis of available physicochemical properties, which can provide some insight into whether the exposure of a candidate molecule may be dissolution rate or solubility limited. *In silico* approaches using biomodeling software such as GastroPlus or Simcyp can be very useful in this respect (Kesisoglou and Wu, 2008; Kuentz, 2008). Multiple examples of decision trees or flowcharts to guide formulation selection on the basis of physicochemical/biopharmaceutical properties have been reported in the literature (Balbach and Korn, 2004; Ku, 2008; Li and Zhao, 2007; Maas *et al.*, 2007; Palucki *et al.*, 2010; Rabinow, 2004), and Figure 2.6 shows one such example developed for use within the authors labs, which shares much in common with those previously reported.

When selecting a formulation strategy to enhance the bioavailability of a poorly soluble compound, it is important to understand the limiting factors for absorption. Drug dissolution in the GIT is generally considered to occur under sink conditions, and if the dissolution rate is slower than the permeation rate across the intestinal epithelium, oral absorption is limited by dissolution. In the framework proposed by Yu (1999), solubility contributes to poor dissolution, but absorption for dissolution rate-limited compounds is strongly influenced by

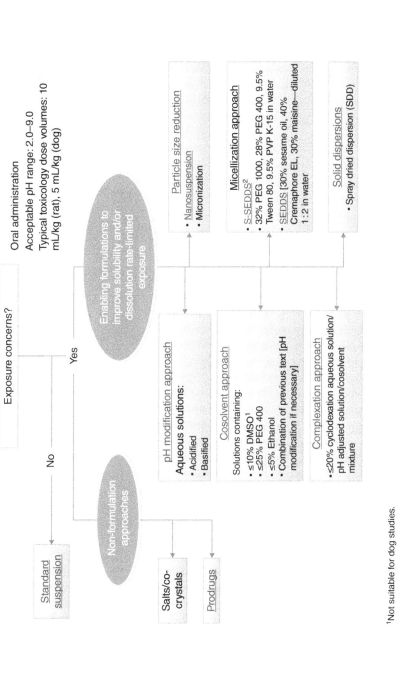

Figure 2.6 Flowchart to guide oral formulation for preclinical *in vivo* studies.

Exposure concerns?

No

Yes

Standard suspension

Non-formulation approaches

Salts/co-crystals

Prodrugs

Oral administration
Acceptable pH range: 2.0–9.0
Typical toxicology dose volumes: 10 mL/kg (rat), 5 mL/kg (dog)

Enabling formulations to improve solubility and/or dissolution rate-limited exposure

pH modification approach
Aqueous solutions:
- Acidified
- Basified

Cosolvent approach
Solutions containing:
- ≤10% DMSO[1]
- ≤25% PEG 400
- ≤5% Ethanol
- Combination of previous text [pH modification if necessary]

Complexation approach
- ≤20% cyclodextran aqueous solution/ pH adjusted solution/cosolvent mixture

Particle size reduction
- Nanosuspension
- Micronization

Micellization approach
- S-SEDDS[2]
- 32% PEG 1000, 28% PEG 400, 9.5% Tween 80, 9.5% PVP K-15 in water
- SEDDS [30% sesame oil, 40% Cremaphore EL, 30% maisine—diluted 1:2 in water

Solid dispersions
- Spray dried dispersion (SDD)

[1]Not suitable for dog studies.
[2]Max dose volume in dog 1mL/kg.

particle size. For such dissolution rate-limited compounds, the absolute amount of drug absorbed increases with increasing dose. For compounds, which are solubility limited (i.e., the dose-to-solubility ratio is high and/or the dissolution rate is far greater than the permeation rate), dissolution occurs under non-sink conditions due to saturation of the solubilizing capacity of the GI lumen. For such solubility-limited compounds, the absolute amount of drug absorbed does not increase with dose, and a plateau in exposure is seen with higher dose levels (Sugano and Terada, 2015; Sugano *et al.*, 2007; Takano *et al.*, 2008). In the context of preclinical studies, it is clearly possible that a compound may be primarily dissolution rate limited at low dose (e.g., screening PK studies) and be solubility limited at higher dose (e.g., for toxicokinetic or toxicology studies). In such a case, it is reasonable to use a simple size reduction strategy to enable early PK studies before contemplating a more complex formulation (SDD, SEDDS, cyclodextrin solubilization) to address the solubility limitation for later toxicology studies.

The selection of an appropriate bio-enhanced approach is usually driven by an analysis of available PK data or predicted absorption profiles from *in silico* biomodeling software, which can provide some insight into whether the exposure of a candidate molecule is likely to be dissolution rate or solubility limited. This information is considered alongside the physicochemical properties of the molecule to select a suitable strategy for bio-enhancement. The biopharmaceutical classification scheme (BCS) is often used to categorize NCEs by virtue of solubility and permeability parameters but can be difficult to apply in a preclinical context as the concept of categorizing on the basis of clinical doses, which are usually unknown or span a very wide range (it is not uncommon for the possible range of clinical doses to span from a few micrograms to a few hundred milligrams at this stage of preclinical development), does not have much relevance to the absorption challenges seen in preclinical species at either PK or toxicological dose levels. However, the general concept of BCS has been adapted by a number of research groups to provide a framework that can be used to inform formulation selection during preclinical and early formulation development. One particular example, termed the "developability classification system" (DCS), provides differentiation to the traditional BCS class II space (low solubility/high permeability) and can be used to refine the selection of suitable formulation approaches based on the biopharmaceutics of the NCE (Butler and Jennifer, 2010). For those compounds, which are deemed to be limited primarily by dissolution rate (DCS IIa), strategies such as particle size reduction by micronization (readily achieved at small scale by bench-top jet-milling equipment) or more aggressive reduction to the nanosized region by wet milling can be effective ways to improve the rate and extent of absorption. For compounds in the DCS IIb region, the solubility limitation predominates, and strategies such as the use of solubilizing systems need to be considered. In terms of technical complexity, the type of solubilizing systems chosen can range from the use of simple cosolvent systems/solubilizing excipients (e.g., low molecular weight PEGS, propylene glycol), cyclodextrin solubilizers, to more sophisticated multicomponent formulation such as SEDDS, SMEDDS, or supersaturable SEDDS (Kawabata *et al.*, 2011). If adequate solubility cannot be achieved in any of these vehicles, it can be possible to exploit the usually higher kinetic solubility of the amorphous form by formulating a stabilized amorphous dispersion, often prepared at small scale by spray-drying an organic solution of API and stabilizing polymer (Friesen *et al.*, 2008).

Many of the challenges associated with development of bio-enhanced pre-clinical formulation development are related to the need for material-sparing, miniaturized processes, which provide reproducible formulations for diverse molecular substrates. In this respect, the use of empirical procedures assisted by computational tools to develop solubilized formulations (by either cosolvents or more complex SEDDS, SNEDDS or SMEDDS approaches) can be readily adapted to high-throughput, multi-well plate-automated screening-type procedures (Gopinathan *et al.*, 2010; Mansky *et al.*, 2007; Sakai *et al.*, 2012). An alternative bio-enhanced formulation approach is to create an amorphous solid dispersion (ASD) in which the drug is molecularly dispersed and stabilized in the amorphous within a polymer matrix. Amorphous dispersions are typically obtained by hot melt extrusion or by spray-drying of a drug/polymer mixture. This can be adapted to small-scale lab production by using a solvent evaporation method where API and hydrophilic polymers are dissolved in a low boiling point organic solvent such as dichloromethane, methanol, ethanol, acetone, isopropanol, ethyl acetate, or a combination of these solvents. Over the last decade the use of ASD has become an increasingly popular strategy to address the solubility challenge. The success of this formulation strategy is reflected in the rise of marketed amorphous products (Smithery *et al.*, 2013; Thayer, 2010). To reduce the complexity of identifying the polymer type, excipients and solvents, and process used for the spray-drying process, a screening methodology can be adopted similar to that published by Duarte *et al.* (2015). With this screening methodology, computer modeling is combined with practical experimentation (solvent casting) that uses small quantities of API to determine the thermodynamic drug solubility in the polymer, optimal drug loading, preferred solvent for spray-drying, and polymer type and ratio, which are required to ensure stability of the formulation during production and long-term storage. Knowledge on the chemical space for application of spray-dried dispersions and associated stability is also well developed (Baumann *et al.*, 2013; Friesen *et al.*, 2008; He and Ho, 2015; Ray, 2012) and facilitates the rapid selection of polymer/drug loading for amorphous system development. The strategy of using an ASD at an early lead development stage can pave the way for onward use into toxicology and possibly clinical formulations. Overall, ASD has been proven to be an effective approach to increase *in vivo* exposure for challenging solubility-limited drugs, and from a practical perspective, the availability of small-scale equipment to provide supplies of dispersions with minimal losses of API is a significant advantage for this technique. A more difficult technique to adapt to small-scale preparation is nanosizing through wet-bead milling processes. While the advantages for nanosizing in terms of enhancing absorption have been clearly demonstrated for a large number of poorly soluble molecules, designing this process to work efficiently and reproducibly with small quantities of drug substance has been challenging for reasons such as process control and loss of yield. A bench-top process suitable for processing small quantities of API and which addresses a number of these concerns is described in the succeeding text (Figures 2.7 and 2.8).

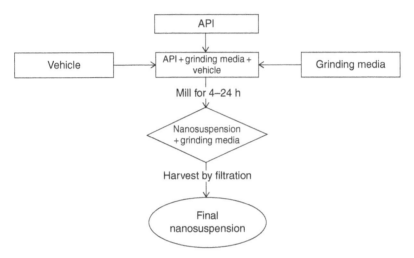

Figure 2.7 Process flow description for bench-top nanosizing method.

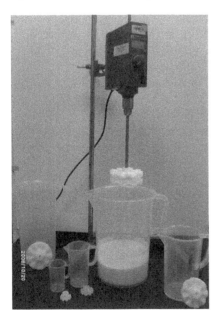

Figure 2.8 Bench-top equipment and impellers used to nanosize API for preclinical formulations.

Results for nanosizing of PF-04191834 (free base with a pK_a of 2.5, essentially nonionizable over the physiological pH range, aqueous solubility ~1 µg/mL, high permeability) are shown in Figure 2.9. It can be seen that the small-scale method described previously that uses simple equipment can effectively reduce micronized API to nanosized material over a 2 h period. No changes in solid state as measured by PXRD were observed, and the crystalline nanosuspension had good chemical and physical stability over a 72 h period, making it suitable for use in both early PK and PD studies.

PK evaluation in a pentagastrin-treated (6 µg/kg via intramuscular injection) dog model showed that both C_{max} and AUC were significantly enhanced by the nanosuspension relative to a micronized aqueous dispersion (Table 2.2).

An alternative technique for preparation of nanoparticles at a small scale is the ContraSol™ emulsion templating developed by IOTA NanoSolutions (IOTA references). This emulsion freeze-drying method uses high internal phase emulsions as templates to create highly porous materials (Cameron, 2005). In contrast, IOTA NanoSolutions™ uses the same emulsion-templating principle and dissolves water-soluble excipients in the aqueous continuous phase of an emulsion while dissolving the drug into the discontinuous oil phase. This emulsion is frozen in a cryogenic liquid to "lock" the emulsion structure and then freeze dried to remove the water and volatile oil phase, leaving a highly porous matrix containing the poorly soluble drug dispersed within (Figure 2.10). This highly porous structure rapidly dissolves in water to release nanoparticles of the drug (Zhang *et al.*, 2008).

As shown in Figure 2.11, this technology was successfully used to nanosize PF-04191834 and illustrates the potential of this technique with its advantages of single-step processing and production of narrow-size distributions to rival attrition-based methods for preclinical nano-formulations.

In preparation of lead candidate selection, most preclinical lead candidates face the challenge of demonstrating acceptable toxicology profile in several species. To achieve this, the high exposures of the candidate to the doses

Table 2.2 PK of PF-04191834 aqueous dispersion and nanosuspension after administration of 100 mg dose to pentagastrin-treated dogs ($n = 5$).

PK parameters	Aqueous dispersion	Crystalline nanosuspension
T_{max} (h)	4 ($\pm$2)	4 ($\pm$2)
C_{max} (ng/mL)	266 ($\pm$91)	1090 ($\pm$27)
$T_{1/2}$ (h)	5 ($\pm$2)	6 ($\pm$2)
AUC_{0-27} (ng h/mL)	4082 ($\pm$2248)	12541 ($\pm$4522)
Bioavailability (%)	9 ($\pm$5)	51 ($\pm$18)

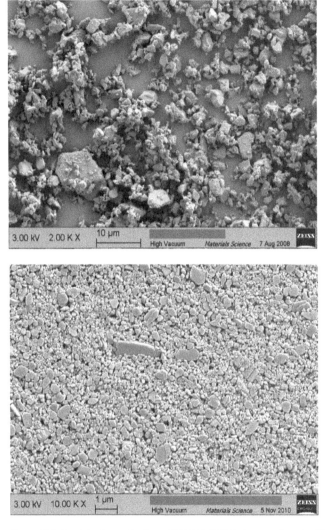

Figure 2.9 Particle size reduction of PF-04191834 after wet-bead milling with bench-top equipment. (*See insert for color representation of the figure.*)

selected in the toxicology study have to be demonstrated, expected, and required. As discussed before, difficulty to achieve adequate oral exposure is often rooted in the suboptimal physicochemical properties (i.e., solubility). As a result, formulation scientists are usually faced with dilemma of choosing tox-friendly enabling formulation strategies to enhance oral absorption of poorly

	Primary key	Product	Batch no.	D[v,0.1] µm	D[v,0.5] µm	D[v,0.9] µm	D[4,3] µm
■	2008-08-13 17:20:29.6620	1257 H PF-04191834	E010008197	0.50	1.78	4.02	2.06
◆	2008-08-13 17:22:10.9580	1257 H PF-04191834	E010008197	0.48	1.74	4.04	2.05
▲	2008-08-13 17:24:34.8170	1257 H PF-04191834	E010008197	0.50	1.75	3.97	2.03

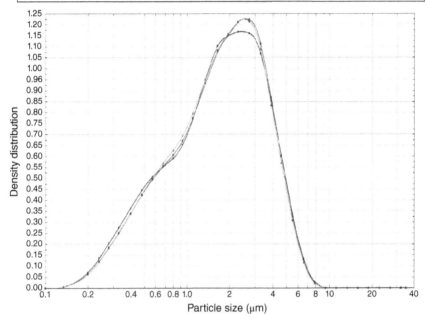

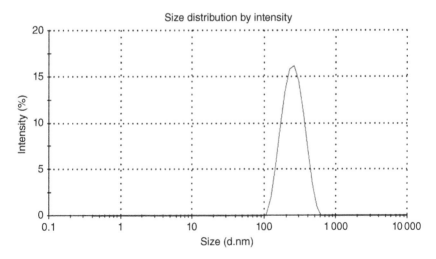

Figure 2.9 (*Continued*)

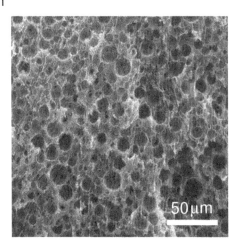

Figure 2.10 Emulsion template structure.

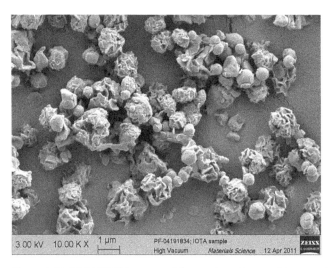

Figure 2.11 PF-04191834 nanoparticles produced by emulsion templating (IOTA Contrasol™ technology).

soluble compounds. With limited supplies of API, it will be challenging to define *in vitro/in vivo* studies that can show exposure at the low dose and at the same time dose proportionality up to the maximum dose defined by the toxicology team. A review of preclinical formulation development strategy that impacts safety assessment studies was recently published by Higgins *et al.* (2012). In this review, besides using a phase formulation approach to achieve exposure multiples, a simple calculation of "maximum absorbable dose" (MAD) as presented by Hilgers *et al.* (2003) was employed to estimate the amount of absorbed drug in GIT using solubility and permeability of the compound. Details of this approach were described by Wuelfing *et al.* (2012) where target compound solubility can be applied for oral absorption at high dose of 10–100 mpk in surfactant suspension or in cosolvent such as PEG 400. This approach was shown to allow for rapid formulation selection using minimal resources.

2.4 Conclusion

The challenges for formulation scientists working to support their medicinal chemistry, biology, and drug metabolism colleagues in early drug discovery teams are very different to those encountered later in development. Formulation strategies at this early stage need to meet aggressive timelines and use minimal quantities of API. The selection of a formulation platform to deliver a lead compound often relies on an incomplete physicochemical and biopharmaceutical profile, and the formulator often resorts to historical experience to guide his or her efforts. However, looking to the future, we can envisage that progress with developing much better informed early biopharmaceutical risk assessments to guide formulation strategy will be driven by the development and adoption of new *in silico* and *in vitro* tools with improved prediction accuracy (Bergström *et al.*, 2014). These tools combined with high-throughput formulation screening technologies will enable early formulation support for drug discovery to transition from a largely empirical process to one that is firmly anchored within a digital formulation design framework.

References

Artursson, P. & Bergström, C. A. S. 2004. Intestinal absorption: the role of polar surface area. In: *Drug Bioavailability: Estimation of Solubility, Permeability, Absorption and Bioavailability* (Han van de, W. Ed.). Wiley-VCH Verlag GmbH & Co. KGaA, Weinheim, pp. 339–357.

Asha, S. & Vidyavathi, M. 2010. Role of human liver microsomes in *in vitro* metabolism of drugs—a review. *Applied Biochemistry and Biotechnology*, 160, 1699–1722.

Avdeef, A. & Testa, B. 2002. Physicochemical profiling in drug research: a brief survey of the state-of-the-art of experimental techniques. *Cellular and Molecular Life Sciences: CMLS*, 59, 1681–1689.

Bajorath, J. 2002. Integration of virtual and high-throughput screening. *Nature Reviews. Drug Discovery*, 1, 882–894.

Balbach, S. & Korn, C. 2004. Pharmaceutical evaluation of early development candidates "the 100 mg approach". *International Journal of Pharmaceutics*, 275, 1–12.

Baumann, J., Dobry, D. & Ray, R. 2013. Amorphous dispersion formulation development: phase-appropriate integrated approaches to optimizing performance, manufacturability, stability & dosage form. *Drug Development & Delivery*, 13, 30–37.

Bergström, C. A., Holm, R., Jørgensen, S. A., Andersson, S. B., Artursson, P., Beato, S., Borde, A., Box, K., Brewster, M. & Dressman, J. 2014. Early pharmaceutical profiling to predict oral drug absorption: current status and unmet needs. *European Journal of Pharmaceutical Sciences*, 57, 173–199.

Bickerton, G. R., Paolini, G. V., Besnard, J., Muresan, S. & Hopkins, A. L. 2012. Quantifying the chemical beauty of drugs. *Nature Chemistry*, 4, 90–98.

Buggins, T. R., Dickinson, P. A. & Taylor, G. 2007. The effects of pharmaceutical excipients on drug disposition. *Advanced Drug Delivery Reviews*, 59, 1482–1503.

Butler, J. M. D. & Jennifer, B. 2010. The developability classification system: application of biopharmaceutics concepts to formulation development. *Journal of Pharmaceutical Sciences*, 99, 4940–4954.

Cameron, N. 2005. High internal phase emulsion templating as a route to well-defined porous polymers. *Polymer*, 46, 1439–1449.

Chaubal, M. V.. 2004. Application of formulation technologies in lead candidate selection and optimization. *Drug Discovery Today*, 9(14), 603–609.

Chen, X.-Q. G., Gudmundsson, O. S. & Hageman, M. J. 2012. Application of lipid-based formulations in drug discovery. *Journal of Medicinal Chemistry*, 55, 7945–7956.

Di, L., Fish, P. V. & Mano, T. 2012. Bridging solubility between drug discovery and development. *Drug Discovery Today*, 17(9–10), 486–495.

Duarte, I., Santos, J. L., Pinto, J. & Temtem, M. 2015. Screening methodologies for the development of spray dried amorphous solid dispersion. *Pharmaceutical Research*, 32, 222–237.

Friesen, D. T., Shanker, R., Crew, M., Smithey, D. T., Curatolo, W. J. & Nightingale, J. A. S. 2008. Hydroxypropyl methylcellulose acetate succinate-based spray-dried dispersions: an overview. *Molecular Pharmaceutics*, 5, 1003–1019.

Gopinathan, S., Nouraldeen, A. & Wilson, A. 2010. Development and application of a high-throughput formulation screening strategy for oral administration in drug discovery. *Future Medicinal Chemistry*, 2, 1391–1398.

Gullapalli, R., Wong, A., Brigham, E., Kwong, G., Wadsworth, A., Willits, C., Quinn, K., Goldbach, E. & Samant, B. 2012. Development of ALZET osmotic pump compatible solvent compositions to solubilize poorly soluble compounds for preclinical studies. *Drug Delivery*, 19, 239–246.

Hann, M. M. 2011. Molecular obesity, potency and other addiction in drug discovery. *Medicinal Chemistry Communications*, 2, 349–355.

He, Y. & Ho, C. 2015. Amorphous solid dispersions: utilization and challenges in drug discovery and development. *Journal of Pharmaceutical Sciences*, 104, 3237–3258.

Hefti, F. F. 2008. Requirements for a lead compound to become a clinical candidate. *BMC Neuroscience*, 9(Suppl 3), S7.

Higgins, J., Cartwright, M. E. & Templeton, A. C. 2012. Foundation review: Progressing preclinical drug candidates: strategies on preclinical safety studies and the quest for adequate exposure. *Drug Discovery Today*, 17(15–16), 828–836.

Hilgers, A. R., Smith, D. P., Biermacher, J. J., Day, J. S., Jensen, J. L., Sims, S. M., Adams, W. J., Fiis, J. M., Palandra, J., Hosley, J. D., Shobe, E. M. & Burton, P. S. 2003. Predicting oral absorption of drugs: a case study with a novel class of antimicrobial agents. *Pharmaceutical Research*, 20, 1149–1155.

Hill, A., Breyer, S., Geissler, S., Mier, W., Haberkorn, U., Weigandt, M. & Maeder, K. 2013. How do *in-vitro* release profiles of nanosuspensions from Alzet pumps correspond to the *in-vivo* situation? A case study on radiolabeled fenofibrate. *Journal of Controlled Release*, 168, 77–87.

Hughes, J. P., Rees, S., Kalindjian, S. B. & Philpott, K. L. 2011. Principles of early drug discovery. *British Journal of Pharmacology*, 162(6), 1239–1249.

Hurst, S., Loi, C.-M., Brodfuehrer, J. & El-Kattan, A. 2007. Impact of physiological, physicochemical and biopharmaceutical factors in absorption and metabolism mechanisms on the drug oral bioavailability of rats and humans. *Expert Opinion on Drug Metabolism & Toxicology*, 3, 469–489.

Kawabata, Y., Wada, K., Nakatani, M., Yamada, S. & Onoue, S. 2011. Formulation design for poorly water-soluble drugs based on biopharmaceutics classification system: basic approaches and practical applications. *International Journal of Pharmaceutics*, 420, 1–10.

Kerns, E. H., Di, L. & Carter, G. T. 2008. *In vitro* solubility assays in drug discovery. *Current Drug Metabolism*, 9, 879–885.

Kesisoglou, F. & Wu, Y. 2008. Understanding the effect of API properties on bioavailability through absorption modeling. *AAPS Journal*, 10, 516–525.

Koltun, M., Morizzi, J., Katneni, K., Charman, S. A., Shackleford, D. M. & Mcintosh, M. P. 2010. Preclinical comparison of intravenous melphalan pharmacokinetics administered in formulations containing either (SBE)7m-β-cyclodextrin or a co-solvent system. *Biopharmaceutics & Drug Disposition*, 31, 450–454.

Ku, S. 2008. Use of the biopharmaceutical classification system in early drug development. *AAPS Journal*, 10, 208–212.

Kuentz, M. 2008. Drug absorption modelling as a tool to define the strategy in clinical formulation development. *AAPS Journal*, 10, 473–479.

Kwong, E., Higgins, J. & Templeton, A. C. 2011. Strategies for bringing drug discovery tools into discovery. *International Journal of Pharmaceutics*, 412, 1–7.

Lee, Y.-C., Zocharski, P. & Samas, B. 2003. An intravenous formulation decision tree for discovery compound formulation development. *International Journal of Pharmaceutics*, 253, 111–119.

Li, P. & Zhao, L. 2007. Developing early formulations: practices and perspective. *International Journal of Pharmaceutics*, 341, 1–19.

Lipinski, C. A., Lombardo, F., Dominy, B. W. & Feeney, P. J. 1997. Experimental and computational approaches to estimate solubility and permeability in drug discovery and development settings. *Advanced Drug Delivery Reviews*, 23, 3–25.

Lipinski, C. A., Lombardo, F., Dominy, B. W. & Feeney, P. J. 2001. Experimental and computational approaches to estimate solubility and permeability in drug discovery and development settings. *Advanced Drug Delivery Reviews*, 46, 3–26.

Maas, J., Kamm, W. & Hauck, G. 2007. An integrated early formulation strategy from hit evaluation to preclinical candidate profiling. *European Journal of Pharmaceutics and Biopharmaceutics*, 66, 1–10.

Manallack, D. T. 2007. The pK(a) distribution of drugs: application to drug discovery. *Perspectives in Medicinal Chemistry*, 1, 25–38.

Mansky, P., Dai, W.-G., Li, S., Pollock-Dove, C., Daehne, K., Dong, L. & Eichenbaum, G. 2007. Screening method to identify preclinical liquid and semi-solid formulations for low solubility compounds: miniaturization and automation of solvent casting and dissolution testing. *Journal of Pharmaceutical Sciences*, 96, 1548–1563.

Murdande, S. B., Pikal, M. J., Shanker, R. M. & Bogner, R. H. 2011. Aqueous solubility of crystalline and amorphous drugs: challenges in measurement. *Pharmaceutical Development and Technology*, 16, 187–200.

Naritomi, Y., Terashita, S., Kagayama, A. & Sugiyama, Y. 2003. Utility of hepatocytes in predicting drug metabolism: comparison of hepatic intrinsic clearance in rats and humans *in vivo* and *in vitro*. *Drug Metabolism and Disposition*, 31, 580–588.

Neervannan, S. 2006. Preclinical formulation for discovery and toxicology: physicochemical challenges. *Expert Opinion on Drug Metabolism & Toxicology*, 2, 715–731.

Niwa, T. & Hashimoto, N. 2008. Novel technology to prepare oral formulations for preclinical safety studies. *International Journal of Pharmaceutics*, 350, 70–78.

Obach, R. S., Baxter, J. G., Liston, T. E., Silber, B. M., Jones, B. C., Macintyre, F., Rance, D. J. & Wastall, P. 1997. The prediction of human pharmacokinetic parameters from preclinical and *in vitro* metabolism data. *Journal of Pharmacology and Experimental Therapeutics*, 283, 46–58.

Palm, K., Luthman, K., Ungell, A.-L., Strandlund, G. & Artursson, P. 1996. Correlation of drug absorption with molecular surface properties. *Journal of Pharmaceutical Sciences*, 85, 32–39.

Palm, K., Stenberg, P., Luthman, K. & Artursson, P. 1997. Polar molecular surface properties predict the intestinal absorption of drugs in humans. *Pharmaceutical Research*, 14, 568–571.

Palucki, M., Higgins, J. D., Kwong, E., Templeton, A. 2010. Strategies at the interface of drug discovery and development: early optimization of the solid state phase and preclinical toxicology formulation for potential drug candidates. *Journal of Medicinal Chemistry*, 53, 5897–5905.

Parkinson, A., Kazmi, F., Buckley, D., Yerino, P., Ogilvie, B. & Paris, B. 2010. System-dependent outcomes during the evaluation of drug candidates as inhibitors of cytochrome P450 (CYP) and uridine diphosphate glucuronosyltransferase (UGT) enzymes: human hepatocytes versus liver microsomes versus recombinant enzymes. *Drug Metabolism and Pharmacokinetics*, 25, 16–27.

Pozarska, A., Da Costa Mathews, C., Wong, M. & Pencheva, K. 2013. Application of COSMO-RS as an excipient ranking tool in early formulation development. *European Journal of Pharmaceutical Sciences*, 49, 505–511.

Rabinow, B. 2004. Nanosuspensions in drug delivery. *Nature Reviews Drug Discovery*, 3, 785–796.

Ray, R. 2012. Addressing solubility challenges: using effective technology and problem-solving for delivery solutions. *Drug Development & Delivery*, 12, 26–29.

Ritter, J. 2007. Intestinal UGTs as potential modifiers of pharmacokinetics and biological responses to drugs and xenobiotics. *Expert Opinion on Drug Metabolism & Toxicology*, 3, 93–107.

Saal, C. & Petereit, A. C. 2012. Optimizing solubility: kinetic versus thermodynamic solubility temptations and risks. *European Journal of Pharmaceutical Sciences*, 47, 589–595.

Sakai, K., Obata, K., Yoshikawa, M., Takano, R., Shibata, M., Maeda, H., Mizutani, A. & Terada, K. 2012. High drug loading self-microemulsifying/micelle formulation: design by high-throughput formulation screening system and *in vivo* evaluation. *Drug Development and Industrial Pharmacy*, 38, 1254–1261.

Saxena, V., Panicucci, R., Joshi, Y. & Garad, S. 2009. Developability assessment in pharmaceutical industry: an integrated group approach for selecting developable candidates. *Journal of Pharmaceutical Sciences*, 98, 1962–1979.

Shah, A. K. & Agnihotri, S. A.. 2011. Recent advances and novel strategies in pre-clinical formulation development: an overview. *Journal of Controlled Release*, 156, 281–296.

Shah, S. M., Jain, A. S., Kaushik, R., Nagarsenker, M. S. & Nerurkar, M. J. 2014. Preclinical formulations: insight, strategies, and practical considerations. *AAPS PharmSciTech*, 15, 1307–1323.

Shore, P. A., Brodie, B. B. & Hogben, C. A. M. 1957. The gastric secretion of drugs: a pH partition hypothesis. *Journal of Pharmacology and Experimental Therapeutics*, 119, 361–369.

Smithery, D., Gao, P. & Taylor, L.. 2013. Amorphous solid dispersions: an enabling formulation technology for oral delivery of poorly water soluble drugs. *AAPS Newsmagazine*, 16(1), 11–14.

Soars, M. G., Burchell, B. & Riley, R. J. 2002. *In vitro* analysis of human drug glucuronidation and prediction of *in vivo* metabolic clearance. *Journal of Pharmacology and Experimental Therapeutics*, 301, 382–390.

Sugano, K. & Terada, K. 2015. Rate- and extent-limiting factors of oral drug absorption: theory and applications. *Journal of Pharmaceutical Sciences*, 104, 2777–2788.

Sugano, K., Kato, T., Suzuki, K., Keiko, K., Sujaku, T. & Mano, T. 2006. High throughput solubility measurement with automated polarized light microscopy analysis. *Journal of Pharmaceutical Sciences*, 95, 2115–2122.

Sugano, K., Okazaki, A., Sugimoto, S., Tavornvipas, S., Omura, A. & Mano, T. 2007. Solubility and dissolution profile assessment in drug discovery. *Drug Metabolism and Pharmacokinetics*, 22, 225–254.

Takano, R., Furumoto, K., Shiraki, K., Takata, N., Hayashi, Y., Aso, Y. & Yamashita, S. 2008. Rate-limiting steps of oral absorption for poorly water-soluble drugs in dogs; prediction from a miniscale dissolution test and a physiologically-based computer simulation. *Pharmaceutical Research*, 25, 2334–2344.

Thayer, A. M.. 2010. Finding solutions: custom manufacturers take on drug solubility issues to help pharmaceutical firms move product through development. *Chemical and Engineering News*, 88(22), 13–18.

Tong, W.-Q. & Whitesell, G. 1999. *In situ* salt-screening—a useful technique for discovery support and preformulation studies. *Pharmaceutical Development and Technology*, 3, 215–223.

Varma, M. V. S., Khandavilli, S., Ashokraj, Y., Jain, A., Dhanikula, A., Sood, A., Thomas, N. S., Pillai, O., Sharma, P., Gandhi, R., Agrawal, S., Nair, V. & Panchagnula, R. 2004. Biopharmaceutic classification system: a scientific framework for pharmacokinetic optimization in drug research. *Current Drug Metabolism*, 5, 375–388.

Wan, H., Holmén, A. G., Wang, Y., Lindberg, W., Englund, M., Någård, M. B. & Thompson, R. A. 2003. High-throughput screening of pKa values of pharmaceuticals by pressure-assisted capillary electrophoresis and mass spectrometry. *Rapid Communications in Mass Spectrometry*, 17, 2639–2648.

Wilson, A. G. E. 2010. A new paradigm for improving oral absorption of drugs in discovery: role of physicochemical properties, different excipients and the pharmaceutical scientists, *Future Medicinal Chemistry*, 2(1), 1–5.

Wuelfing, W. P., Kwong, E. & Higgins, J.. 2012. Identification of suitable formulations for high dose oral studies in rats using *in vitro* solubility measurements, the maximum absorbable dose model, and historical data sets. *Molecular Pharmaceutics*, 9, 1163–1174.

Yamagami, C., Kawase, K. & Iwaki, K. 2002. Hydrophobicity parameters determined by reversed phase liquid chromatography. XV: optional conditions for prediction of log P(oct) by using RP-HPLC procedures. *Chemical & Pharmaceutical Bulletin*, 50, 1578–1583.

Yamashita, S., Furubayashi, T., Kataoka, M., Sakane, T., Sezaki, H. & Tokuda, H. 2000. Optimized conditions for prediction of intestinal drug permeability using Caco-2 cells. *European Journal of Pharmaceutical Sciences*, 10, 195–204.

Yu, L. 1999. An integrated model for determining causes of poor oral drug absorption. *Pharmaceutical Research*, 16, 1883–1887.

Zhang, H., Wang, D., Butler, R., Campbell, N. L., Long, J., Tan, B., Duncalf, D. J., Foster, A. J., Hopkinson, A., Taylor, D., Angus, D., Cooper, A. I. & Rannard, S. P. 2008. Formation and enhanced biocidal activity of water-dispersible organic nanoparticles. *Nature Nanotechnology*, 3, 506–511.

3

Oral Drug Formulation Development in Pharmaceutical Lead Selection Stage

Shayne Cox Gad

Gad Consulting Services, Cary, NC, USA

3.1   Introduction

Drug discovery lead optimization is usually a stage wherein the search for a good preclinical candidate is selected. At this stage, a good balancing act of optimizing potency, pharmacokinetics (PK) behavior and pharmacodynamics (PD), and toxicity is performed. To explore acceptable toxicology profile, the

Oral Formulation Roadmap from Early Drug Discovery to Development,
First Edition. Edited by Elizabeth Kwong.
© 2017 John Wiley & Sons, Inc. Published 2017 by John Wiley & Sons, Inc.

lead had to meet high exposures. It is in the early stages of screening and lead selection toxicology that one must start to have formulations achieve their objective for full development: to deliver the maximum possible (and tolerable) dose of the candidate drug(s) in the lowest volume of formulation while maintaining the stability of the active drug moiety and while optimizing systemic absorption in the model species used.

In recent years, the process of a new drug taken from concept to clinic involves several steps that occurs in more efficient strategies to screen, identify, and optimize lead compounds for development. High-throughput screening and computational methodologies have been used (Kwong *et al.*, 2011). However, use of simple conventional formulation technologies is therefore not suitable for compounds resulting as "hits" from these screening processes, because of poor aqueous solubility, compound availability, and poorly characterized materials.

Before the development of an optimized formulation, however, preclinical evaluations of the safety of the drug moiety must be performed and clinical and preclinical formulations have slightly different requirements, and usually preclinical formulations have a wider range of potential vehicles to use. In most instances, rapid formulation development such as pH adjustment and *in situ* salt formation depending on structure are mostly used at this stage to improve the formulation of series that are exhibiting non-wettability during formulation. Currently, novel-enabling technologies such as solid dispersions, which increased active pharmaceutical ingredients (API) demand by 10-fold, are usually reserved at a later stage of lead optimization. Although challenging, formulation approaches using amorphous solid dispersions in discovery were exemplified by Kwong *et al.* (2011) recognizing (i) the limited API supplies, (ii) time constraints, (iii) physical stability of amorphous form, and (iv) maximization of exposure up to 750 mpk.

3.2 Formulation Considerations in Lead Selection Stage

The perfect drug would be one along the lines of Paul Ehrlich's "Magic Bullet"— as illustrated in Figure 3.1, a drug molecule is readily administered, is completely absorbed, moves to the desired therapeutic target site (protein receptor), does what is supposed to, and is completely eliminated from the body. Complete therapeutic target specificity, the most pressing (and rewarding, if successful) area for drug development currently, is optimizing the drug to therapeutic target delivery part of this process. Once the intended route of administration is selected (Table 3.1 presents all of the common potential choices), successively more sophisticated formulations must be developed. One of the key steps in the nonclinical and clinical formulation of the drug is the selection of vehicles

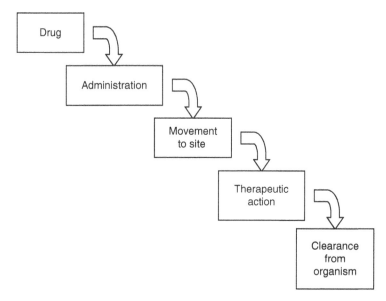

Figure 3.1 The magic bullet concept.

and of the inactive ingredients (excipients). Excipients are essential components of drug products (DPs) in the United States, and one must adequately address the safety of the proposed exposure to the excipients in those products. The specific safety data that may be needed will vary depending upon the clinical situation, including such factors as the duration, level, and route of exposure (i.e., means of clinical drug administration). Which formulation components are acceptable will change over the course of development and in accordance with which species is to receive the test formulation.

Many guidances exist to aid in the development of pharmaceutical drugs, but very few guidances exist to aid in the formulation of drugs for nonclinical safety evaluation, or for the assessment of pharmaceutical excipient safety. The Food and Drug Administration (FDA)/Center for Drug Evaluation and Research (CDER) adopted, in 2005, the guidance for industry "Nonclinical Studies for Development of Pharmaceutical Excipients," which focuses on the development of safety profiles to support use of new excipients as components of drug or biological products.

Similar guidance was published by International Pharmaceutical Excipient Council (IPEC) association "Excipient Safety Evaluation Guidance" in 1995 (updated in 2012 (IPEC, 2012)). These guidelines are presented in a tiered approach of recommended data that should be available on an excipient to provide a pharmaceutical formulator with a rational basis for including a new excipient in a drug formulation.

Table 3.1 Potential routes of administration.

A) Oral routes
    1) Oral (PO)[a]
    2) Inhalation[a]
    3) Sublingual
    4) Buccal
B) Placed into a natural orifice in the body other than the mouth
    1) Intranasal
    2) Intra-auricular
    3) Rectal
    4) Intrafaginal
    5) Intrauterine
    6) Intraurethral
C) Parenteral (injected into the body or placed under the skin)
    1) Intravenous (IV)[a]
    2) Subcutaneous (SC)[a]
    3) Intramuscular (IM)[a]
    4) Intra-arterial
    5) Intradermal (ID)[a]
    6) Intralesional
    7) Epidural
    8) Intrathecal
    9) Intracisternal
    10) Intracardiac
    11) Intraventricular
    12) Intraocular
    13) Intraperitoneal (IP)[a]
D) Topical routes
    1) Cutaneous[a]
    2) Transdermal (also called percutaneous)[a]
    3) Ophthalmic[a]

[a] Commonly used in safety assessment.

The three essential requirements of API principles are compared with those of excipients. Fundamental for both are quality and safety. The requirement of therapeutic efficacy for drugs is replaced by that of functionality for excipient, defined as "the physical, physicochemical, and biopharmaceutical properties" of the same physicochemical factors (e.g., pH and osmolality).

Throughout the development process for pharmaceuticals, formulation development is proceeding with several objectives in mind. The importance of each of these factors changes over time (Monkhouse and Rhodes, 1998) and as

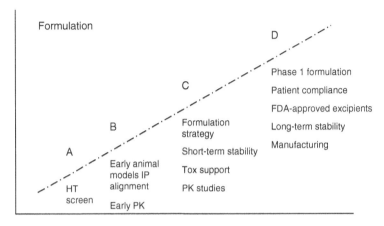

Figure 3.2 Evolution of formulations through phase 1.

illustrated in Figure 3.2. First is optimizing the bioavailability of the therapeutic target organ site by the intended clinical route. Clinical route(s) is selected on a number of grounds (nature of the drug, patient acceptance, issues of safety). Second is minimizing any safety concerns. This means not only systemic toxicity but also local tissue tolerance at the site of application. Third is optimizing stability of the drug active ingredient. Its activity and integrity must be maintained for long enough to be made effectively available to patients. Early on in pre-clinical development, simplicity and maximized bioavailability are essential. Early single dose studies in animals are the starting place and usually bear no relationship to what is used later (Groves, 1966).

More recently, several publications had reviewed research and development (R&D) productivity. Reduction of late phase attrition by assessing physicochemical properties relative to its effect on exposure that contributed to proper assessment of toxicity and potency is highly recommended (Higgins *et al.*, 2012; Lohani *et al.*, 2014; Palucki *et al.*, 2010). Properties such as solubility, lipophilicity, permeability, physical form, physical, and chemical stability are now routinely incorporated into the drug-like properties. Additionally, identifying associated solid-state phase that has appropriate physicochemical characteristics would allow for optimal *in vivo* performance. Compound physical form such as salt, crystalline, or amorphous can profoundly affect solubility and robustness of its formulation. Strategies to interface preformulation at the discovery stage including a concise workflow were described by Palucki *et al.* (2010), Caldwell *et al.* (2001), and Lohani *et al.* (2014). Challenges of API availability and examples driving success of the multidisciplinary approach to lead optimization were also thoroughly explained: from early demonstration of dose-limiting toxicity to early identification of optimal phase and formulation, all contributed to lowering impact failure of a molecule to market. Finally, Hageman (2010) also described the key physicochemical properties that impact "developability" of the selected candidate.

3.2.1 Preformulation

While advances in molecular biology and genomics have produced a flood of molecules with vastly improved target receptor specificity, these molecules have frequently turned out to be very difficult to get absorbed and to the desired target tissue site. Preformulation is an effort to understand the physicochemical (and solvent interaction) aspects of a drug molecule to allow a more effective approach to then developing a formulation. There is no single factor that can account for absorption and formulatability of an NCE. Factors such as lipophilicity, solubility, pK_a, absorption, metabolism, and pharmacokinetics are all interrelated. Therefore it is important to recognize the preformulation as an integral part of lead selection (Adeyeye and Brittain, 2008; Hageman, 2010).

Lipinski's rule of five (RO5, Lipinski, 2004; Lipinski *et al.*, 2001) predates these recent target advances in specificity, but not the problems. In its original form, the RO5 proposed four guiding principles:

- No more than five hydrogen bond donors (the total number of nitrogen–hydrogen and oxygen–hydrogen bonds)
- Not more than 10 hydrogen bond acceptors (all nitrogen or oxygen atoms)
- A molecular mass less than 500 Da
- An octanol–water partition coefficient log$P$ not greater than 5

While Lipinski *prima facie* applies to oral route drugs, it also is useful for other routes.

For clinically useful drugs (and therefore for drugs proceeding through preclinical and nonclinical evaluation and development), there are a number of desirable attributes:

- Ionization at physiologically relevant pH
- Adequate solubility at biorelevant media
- Permeable
- A simple structure
- Simple and efficient synthesis
- Nonhygroscopic
- Avoid chiral centers
- Lack of mutagenicity
- Crystalline with solid-state stability
- No strong odors, colors, or (of oral) tastes
- Compatible with standard excipient
- Physically and chemically stable at ambient temperatures and at physiologic pHs (Adeyeye and Brittain, 2008; Gibson, 2009; Niazi, 2007).

3.2.2 Exposure in Preclinical Species

Bioavailability is defined as the fraction of the dose reaching either the therapeutic target organ or tissue or the systemic circulation as unchanged compound following administration by any route. For an agent administered orally, bioavailability may be less than unity, for several reasons. The molecule may be incompletely absorbed. It may be metabolized in the gut, the gut wall, the portal blood, or the liver prior to entry into the systemic circulation (see Figure 3.3). It may undergo enterohepatic cycling with incomplete reabsorption following elimination into the bile. Biotransformation of some chemicals in the liver following oral administration is an important factor in the pharmacokinetic profile, as will be discussed further. Bioavailability measures following oral administration are generally given as the percentage of the dose available to the systemic circulation.

As the components of a mixture may have various physiochemical characteristics (solubility, vapor pressure, density, etc.), great care must be taken in preparing and administering any mixture so that what is actually tested is the mixture of interest.

Examples of such procedures are making dilutions (not all components of the mixture may be equally soluble or miscible with the vehicle) and generating either vapors or respirable aerosols (not all the components may have equivalent volatility or surface tension, leading to a test atmosphere that contains only a portion of the components of the mixture).

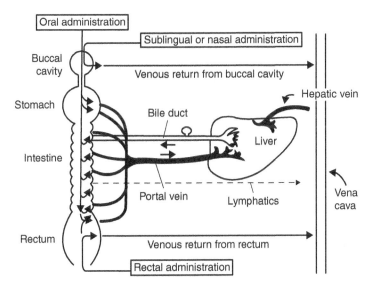

Figure 3.3 Path of drugs through the body after absorption by one of the three routes of administration.

By increasing or decreasing the viscosity of a formulation, the absorption of a toxicant can be altered (Ritschel *et al.*, 1974). Conversely, the use of absorbent to diminish absorption has been used as an antidote therapy for some forms of intoxication. Using the knowledge that rats cannot vomit, there have been serious attempts at making rodenticides safer to nontarget animals by incorporating emetics into the formulations, but this has had only a limited success. Gaines used *in vivo* liver perfusion techniques to investigate the apparent anomaly that the carbamate Isolan was more toxic when administered to rats percutaneously than when administered orally (Gaines, 1960). It has been shown that these results, a manifestation of different formulations, have been used for the two routes of exposure (oral and percutaneous) in estimating the LD_{50} values using a common solvent, *n*-octanol. It was found that Isolan was significantly more toxic by the oral route than by the percutaneous route; by regression analysis it was found that at no level of lethal dose values was the reverse correct.

Although the oral route is the most convenient for most therapeutic uses, there are numerous factors that make it unpredictable, particularly for drug molecules that have very limited water solubility (Liu, 2008). Absorption by this route is subject to significant variation from animal to animal and even in the same individual animal at different times. Considerable effort has been spent by the pharmaceutical industry to develop drug formulations with absorption characteristics that are both effective and dependable. Protective enteric coatings for pharmaceuticals were introduced long ago to retard the action of gastric fluids and then disintegrate and dissolve after passage of a tablet into the human intestine. The purposes of these coatings for drugs are to protect the active ingredient, which would be degraded in the stomach, to prevent nausea and vomiting caused by local gastric irritation (also a big problem in rodent studies, where over a long time period gastric irritation frequently leads to forestomach hyperplasia), to obtain higher local concentrations of the active ingredient intended to act locally in the intestinal tract, to produce a delayed biological effect, or to deliver the active ingredient to the intestinal tract for optimal absorption there. Such coatings are generally fats, fatty acids, waxes, or other such agents, and all of these intended purposes for drug delivery can readily be made to apply for some toxicity studies. Their major drawback, however, is the marked variability in time for a substance to be passed through the stomach. In humans, this gastric emptying time can range from minutes to as long as 12 h. One would expect the same for animals, as the limited available data suggest is the case. Similar coating systems, including microencapsulation (see Melnick *et al.*, 1987), are available for, and currently used in, animal toxicity studies.

The test chemical is unlikely to be absorbed or excreted unless it is first released from its formulation. It is this stage of the process that is the first and most critical step for the activity of many chemicals. If the formulation does not release the chemical, the rest of the process becomes somewhat pointless.

It might be argued that the simplest way around the formulation problem is to administer any test as a solution in water, thereby avoiding the difficulties altogether. However, since multiple, small, accurately measured doses of a chemical are required repeatedly, reproducible dilutions must be used. Also, the water itself is to be regarded as the formulation vehicle, and the test substance must be water soluble and stable in solution, which many are not. The problem can become complex, if we take into account the needs for accuracy, stability, and optimum performance *in vivo*.

Direct connections between observed toxicity and formulation components are uncommon, and it is usually assumed that vehicles and other non-test chemical components are innocuous or have only transitory pharmacological effects. Historically, however, this has certainly not been the case. Even lactose may have marked toxicity in individual test animals (or humans) who are genetically incapable of tolerating it.

The initial stage of drug release from the formulation, both in terms of the amount and the rate of release, may exercise considerable influence at the clinical response level. A close consideration of the formulation parameters of any chemical is therefore essential during the development of any new drug, and, indeed, there are examples where formulations of established drugs also appear to require additional investigation. In addition, it is imperative that the excipients not have any effect on the pharmacodynamic response. Several excipients are known to interact with biological functions such as inhibition of P-glycoprotein (Batrakova *et al.*, 1999; Gough *et al.*, 1982) and intestinal CYP3A4 enzymes (Mountfield *et al.*, 2000; Van Zuylen *et al.*, 2001; Wagner *et al.*, 2001).

The effects of formulation additives on chemical bioavailability from oral solutions and suspensions have been previously reviewed by Swarbrick (2006). He pointed out how the presence of sugars in a formulation may increase the viscosity of the vehicle. However, sugar solutions alone may delay stomach-emptying time considerably when compared with solutions of the same viscosity prepared with celluloses, which may be due to sugar's effect on osmotic pressure. Sugars of different types may also have an effect on fluid uptake by tissues, and this, in turn, correlates with the effect of sugars such as glucose and mannitol on drug transport. It is also known that large caloric loads and large lipid loads would prolong stomach-emptying rates and subsequently gastrointestinal (GI) transit times. Schulze *et al.* (2003) mentioned that PEG 400 can influence GI transport and, depending on the concentration of PEG 400, can either facilitate drug absorption at low PEG or decrease absorption at higher dose due to increased transit time. Therefore it is traditionally common to consider doing abbreviated profiles by looking at the rate of absorption, C_{max}, T_{max}, and terminal half-life, using these formulations to see if there are any effects due to the combination of the formulation and test compound.

Surfactants have been explored widely for their effects on drug absorption, in particular using experimental animals (Gibaldi, 1976; Gibaldi and Feldman, 1970). Surfactants alter dissolution rates (of lipid materials), surface areas of particles and droplets, and membrane characteristics, all of which affect absorption.

Surfactants may increase the solubility of the drug via micelle formation, but the amounts of material required to increase solubility significantly are such that at least orally the laxative effects are likely to be unacceptable and in some species (the dog) there are immediate adverse innate immune responses associated with surfactant use. The competition between the surfactant micelles and the absorption sites is also likely to reduce any useful effect and make any prediction of net overall effect difficult. However, if a surfactant has any effect at all, it is likely to be in the realm of agents that help disperse suspensions of insoluble materials and make them available for solution. Natural surfactants, in particular bile salts, may enhance absorption of poorly soluble materials.

The effective surface area of an ingested chemical is usually much smaller than the specific surface area that is an idealized *in vitro* measurement. Many drugs whose dissolution characteristics could be improved by particle size reduction are extremely hydrophobic and may resist wetting by GI fluids. Therefore, the GI fluids may come in intimate contact with only a fraction of the potentially available surface area. The effective surface area of hydrophobic particle can often be increased by the addition of a surface-active agent to the formulation, which reduces the contact angle between the solid and the GI fluids, thereby increasing effective surface area and dissolution rate.

Disregarding such chemical-specific properties as dissociation constants (in the case of ionic compounds), particle size, and polymorphism, as well as side effects of viscosity, binding to vehicle components, complex formulation, and the like, the following formulation principles arise:

1) Optimization of the concentration of chemical capable of diffusion by testing its maximum solubility
2) Reduction of the proportion of solvent to a degree that is adequate to keep the test material still in solution
3) Use of vehicle components that reduce the permeability barriers

These principles lead to the conclusion that each test substance requires an individual formulation. Sometimes different ingredients will be required for different concentrations to obtain the maximum rate of release. No universal vehicle is available for any route, but a number of primary approaches are. Any dosage preparation lab should be equipped with glassware, a stirring hot plate, a sonicator, a good homogenizer, and a stock of the basic formulating material, as detailed at the end of this chapter (Strickley, 2008, 2011).

Current publications (e.g., Palucki *et al.*, 2010; Shah and Agnihotri, 2011; Zheng *et al.*, 2012) emphasized best practices in pharmaceutical industry where biorelevant *in vitro* screening coupled with *in silico* modeling helps design better structure–activity relationships (SAR). More recently, a commentary (Kwong, 2015) was also published that showed some critical information that could complement the planning of the formulation selection for oral nonclinical safety studies. In summary, the use of physical chemistry and principles of formulation during lead optimization will result in candidates wherein the drug delivery is built into the molecule and reduces the complexity of formulation in preclinical and clinical stage.

3.2.3 Mechanisms of Absorption

There are three primary sets of reasons why differences in formulations and the route of administration are critical in determining the effect of an agent of the biological system. These are (i) local effects, (ii) absorption and distribution, and (iii) metabolism:

Local effects
Local effects are those that are peculiar to the first area or region of the body to which a test material gains entry or that it contacts. For the oral route, these include irritation, corrosion, emesis, and potential innate immune response. In general, the same categories of possible adverse effects (irritation, immediate immune response, local tissue–cellular compatibility, and physicochemical interactions) are the mechanisms of, or basis for, concern. Attention to the viscosity, pH, and osmolality of a formulation can limit these.

In general, no matter what the route, certain characteristics will predispose a material to have local effects (and, by definition, if not present, tend to limit the possibility of local effects). These factors include pH, redox potential, high molar concentration, and low flexibility and sharp edges of certain solids. These characteristics will increase the potential for irritation by any route and, subsequent to the initial irritation, other appropriate regional adaptive responses (for orally administered materials, e.g., emesis and diarrhea).

Absorption and distribution
For a material to be toxic, it must be absorbed into the organism (local effects are largely not true toxicities by this definition).

There are characteristics that influence absorption by the different routes, and these need to be understood by any person trying to evaluate and/or predict the toxicities of different moieties. Some key characteristics and considerations are summarized in the following sections by route.

Table 3.2 presents the normal pH ranges for human physiological fluids. These need to be considered in terms of the impact on solubility and stability of a formulation and active drug.

Table 3.2 Normal pH range for human physiologic fluids.

Medium	Normal pH range
Tears	7.35–7.45
Saliva	6.0–8.0
Gastric juice	1.5–6.5
Intestinal juice	6.5–7.6
Blood	7.35–7.45
Skin (sweat)	4.0–6.8

Metabolism

Metabolism is directly influenced both by the region a material is initially absorbed into and by distribution (both the rate and the pattern). Rate determines whether the primary enzyme systems will handle the entire xenobiotic dose or these are overwhelmed. Pattern determines which routes of metabolism are operative.

3.2.3.1   Oral Routes (Direct to GI Tract)

1) Lipid-soluble compounds (nonionized) are more readily absorbed than water-soluble compounds (ionized):
   a) Weak organic basses are in the nonionized, lipid-soluble form in the intestine and tend to be absorbed there.
   b) Weak organic acids are in the nonionized, lipid-soluble form in the stomach, and one would suspect that they would be absorbed there, but absorption in the intestine is greater because of time and area of exposure.
2) Specialized transport systems exist for some moieties: sugars, amino acids, pyrimidines, calcium, and sodium.
3) Almost everything is absorbed—at least to a small extent (if it has a molecular weight below 10 000).
4) Digestive fluids may modify the structure of a chemical.
5) Dilution increases toxicity because of more rapid absorption from the intestine, unless stomach contents bind the moiety.
6) Physical properties are important; for example, dissolution of metallic mercury is essential to allow its absorption.
7) Age is important; for example, neonates have a poor intestinal barrier.
8) Effect of fasting on absorption depends on the properties of the chemical of interest.

While our concern in this volume is the oral route, as a generalization, there is a pattern of relative absorption rates that characterize the different routes that are commonly employed. This order of absorption by rate from fastest to slowest and, in a less rigorous manner, by degree of absorption from most to least is IV > inhalation > IM > IP > SC > oral > ID > other.

Exposure (total amount and rate, distribution, metabolism) and species similarity in response are the reasons for selecting particular routes in toxicology in general. In safety assessment of pharmaceuticals, however, the route is usually dictated by the intended clinical route and dosing regimen. If this route of human exposure is uncertain or if there is the potential for either a number of routes of the human absorption rate and pattern being greater, then the common practice becomes that of the most conservative approach. This approach stresses maximizing potential absorption in the animal species (within the limits of practicality) and selecting from among those routes commonly used in the laboratory that get the most material into the animal's system as quickly and completely as possible to evaluate the potential toxicity. Under this approach, many compounds are administered intraperitoneally in acute testing, though there is little or no real potential for human exposure by this route.

Assuming that a material is absorbed, distribution of a compound in early preclinical studies is usually of limited interest. In so-called heavy acute studies (Gad *et al.*, 1984) where acute systemic toxicity is intensive and evaluated to the point of identifying target organs, or in range-finder-type study results, for refining the design of longer-term studies, distribution would be of interest. Some factors that alter distribution are listed in Table 3.3.

The oral route is the most commonly used route for the administration of drugs both because of ease of administration and because it is the most readily accepted route of administration. Although the dermal route may be as common for occupational exposure, it is much easier to accurately measure and administer doses by the oral route.

Enteral routes technically include any that will put a material directly into the GI tract, but the use of enteral routes other than oral (such as rectal) is rare in toxicology. Though there are a number of variations of technique and peculiarities of animal response that are specific to different animal species, there is also a great deal of commonality across species in methods, considerations, and mechanisms.

3.2.3.2 Absorption

Ingestion is generally referred to as oral or peroral (PO) exposure and includes direct intragastric exposure in experimental toxicology. The regions for possible agent action and absorption from PO absorption should, however, be considered separately.

Because of the rich blood supply to the mucous membranes of the mouth (buccal cavity), many compounds can be absorbed through them. Absorption from the buccal cavity is limited to nonionized, lipid-soluble compounds. Buccal absorption of a wide range of aromatic and aliphatic acids and basic drugs in human subjects has been found to be parabolically dependent on log$P$,

Table 3.3 Test subject characteristics that can influence GI tract absorption.[a]

A) General and inherent characteristics
   1) General condition of the subject (e.g., starved versus well-fed, ambulatory vs. supine)
   2) Presence of concurrent diseases (i.e., diseases may either speed or slow gastric emptying)
   3) Age
   4) Weight and degree of obesity
B) Physiological function
   1) Status of the subject's renal function
   2) Status of the subject's hepatic function
   3) Status of the subject's cardiovascular system
   4) Status of the subject's gastrointestinal motility and function (e.g., ability to swallow)
   5) pH of the gastric fluid (e.g., affected by fasting, disease, food intake, drugs)
   6) Gastrointestinal blood flow to the area of absorption
   7) Blood flow to areas of absorption for dose forms other than those absorbed through gastrointestinal routes
C) Acquired characteristics
   1) Status of the subject's anatomy (e.g., previous surgery)
   2) Status of the subject's gastrointestinal flora
   3) Timing of drug administration relative to meals (i.e., presence of food in the gastrointestinal tract)
   4) Body position of subject (e.g., lying on the side slows gastric emptying)
   5) Psychological state of subject (e.g., stress increases gastric emptying rate and depression decreases rate)
   6) Physical exercise of subject may reduce gastric emptying rate
D) Physiological principles
   1) Food enhances gastric blood flow, which should theoretically increase the rate of absorption
   2) Food slows the rate of gastric emptying, which should theoretically slow the rate of passage to the intestines where the largest amounts of most agents are absorbed. This should decrease the rate of absorption for most agents. Agents absorbed to a larger extent in the stomach will have increased time for absorption in the presence of food and should be absorbed more completely than in fasted patients
   3) Bile flow and secretion are stimulated by fats and certain other foods. Bile salts may enhance or delay absorption depending on whether they form insoluble complexes with drugs or enhance the solubility of agents
   4) Changes in splanchnic blood flow as a result of food depend on direction and magnitude of the type of food ingested
   5) Presence of active (saturable) transport mechanisms places a limit on the amount of a chemical that may be absorbed

---

[a] The minimization of variability due to these factors rests on the selection of an appropriate animal model, careful selection of healthy animals, and use of proper techniques.

where P is the octanol–water partition coefficient. The ideal lipophilic character ($\log P_0$) for maximum buccal absorption has also been shown to be in the range 4.2–5.5 (Bates and Gibaldi, 1970; Lien *et al.*, 1971). Compounds with large molecular weights are poorly absorbed in the buccal cavity, and, since absorption increases linearly with concentration and there is generally no difference between optical enantiomorphs of several compounds known to be absorbed from the mouth, it is believed that uptake is by passive diffusion rather than by active transport chemical moieties.

A knowledge of the buccal absorption characteristics of a chemical can be important in a case of accidental poisoning. Although an agent taken into the mouth will be voided immediately on being found objectionable, it is possible that significant absorption can occur before any material is swallowed.

Unless voided, most materials in the buccal cavity are swallowed. No significant absorption occurs in the esophagus and the agent passes on to enter the stomach. It is common practice in safety assessment studies to avoid the possibility of buccal absorption by intubation (gavage) or by administration of the agent in gelatin capsules designed to disintegrate in the gastric fluid.

Absorption of chemicals with widely differing characteristics can occur at different levels in the GI tract (Schranker, 1960). The two factors primarily influencing this regional absorption are (i) the lipid–water partition characteristics of the undissociated toxicant and (ii) the dissociation constant (pK_a) that determines the amount of toxicant in the dissociated form.

Therefore, weak organic acids and bases are readily absorbed as uncharged lipid-soluble molecules, whereas ionized compounds are absorbed only with difficulty, and nonionized toxicants with poor lipid-solubility characteristics are absorbed slowly. Lipid-soluble acid molecules can be absorbed efficiently through the gastric mucosa, but bases are not absorbed in the stomach.

In the intestines the nonionized form of the drug is preferentially absorbed, and the rate of absorption is related to the lipid–water partition coefficient of the toxicant. The highest pK_a value for a base compatible with efficient gastric absorption is about 7.8, and the lowest pK_a for an acid is about 3.0, although a limited amount of absorption can occur outside these ranges. The gastric absorption and the intestinal absorption of a series of compounds with different carbon chain lengths follow two different patterns. Absorption from the stomach increases as the chain lengthens from methyl to *n*-hexyl, whereas intestinal absorption increases over the range methyl to *n*-butyl and then diminishes as the chain length further increases. Houston *et al.* (1974) concluded that to explain the logic of optimal partition coefficients for intestinal absorption, it was necessary to postulate a two-compartment model with a hydrophilic barrier and a lipoidal membrane and that if there is an acceptable optimal partition coefficient for gastric absorption, it must be at least ten times greater than the corresponding intestinal value.

Because they are crucial to the course of an organism's response, the rate and extent of absorption of biologically active agents from the GI tract also have major implications for the formulation of test material dosages and also for how production (commercial) materials may be formulated to minimize potential accidental intoxications while maximizing the therapeutic profile.

There are a number of separate mechanisms involved in absorption from the GI tract:

Passive absorption. The membrane lining of the tract has a passive role in absorption. As toxicant molecules move from the bulk water phase of the intestinal contents into the epithelial cells, they must pass through two membranes in series: one layer of water and the other the lipid membrane of the microvilli surface (Wilson and Dietschy, 1974). The water layer may be the rate-limiting factor for passive absorption into the intestinal mucosa, but it is not rate-limiting for active absorption. The concentration gradient and the physicochemical properties of the drug and of the lining membrane are the controlling factors. Chemicals that are highly lipid soluble are capable of passive diffusion, and they pass readily form the aqueous fluids of the gut lumen through the lipid barrier of the intestinal wall and into the bloodstream. The interference in the absorption process by the water layer increases with increasing absorbability of the substances in the intestine (Winne, 1978).

Aliphatic carbamates are rapidly absorbed from the colon by passive uptake (Wood *et al.*, 1978), and it is found that there is a linear relationship between $\log K_a$ and $\log P$ for absorption of these carbamates in the colon and the stomach, whereas there is a parabolic relationship between these two values for absorption in the small intestine. The factors to be considered are:

P = octanol–buffer partition coefficient
K_a = absorption rate constant
t = time
$t_{1/2}$ = half-life = $\ln 2/K_a$

Organic acids that are extensively ionized at intestinal pHs are absorbed primarily by simple diffusion.

Facilitated diffusion. Temporary combination of the chemical with some form of "carrier" occurs in the gut wall, facilitating the transfer of the toxicant across the membranes. This process is also dependent on the concentration gradient across the membrane, and there is no energy utilization in making the translocation. In some intoxications, the carrier may become saturated, making this the rate-limiting step in the absorption process.

Active transport. As discussed later, the process depends on a carrier but differs in that the carrier provides energy for translocation from regions of lower concentration to regions of higher concentration.

Pinocytosis. This process by which particles are absorbed can be an important factor in the ingestion of particulate formulations of chemicals (e.g., dust formulations, suspensions of wettable powders, etc.); however, it must not be confused with absorption by one of the aforementioned processes, where the agent has been released from particles.

Absorption via lymphatic channels. Some lipophilic chemicals dissolved in lipids may be absorbed through the lymphatics.

Convective absorption. Compounds with molecular radii of less than 4 nm can pass through pores in the gut membrane. The membrane exhibits a molecular sieving effect.

Characteristically, within certain concentration limits, if a chemical is absorbed by passive diffusion, then the concentration of toxicant in the gut and the rate of absorption are linearly related. However, if absorption is mediated by active transport, the relationship between concentration and rate of absorption conforms to Michaelis–Menten kinetics and Lineweaver–Burk plot (i.e., reciprocal of rate of absorption plotted against reciprocal of concentration), which graphs as a straight line.

Differences in the physiological chemistry of GI fluids can have a significant effect on toxicity. Both physical and chemical differences in the GI tract can lead to species differences in susceptibility to acute intoxication. The antihelminthic pyrvinium chloride has an identical LD_{50} value when administered intraperitoneally to rats and mice ($\sim$4 mg/kg); when administered orally, however, the LD_{50} value in mice was found to be 15 mg/kg, while for the rat, the LD_{50} values were 430 mg/kg for females and 1550 mg/kg for males. It is thought that this is an absorption difference rather than a metabolic difference (Ritschel *et al.*, 1974).

Most of any exogenous chemical absorbed from the GI tract must pass through the liver via the hepatic portal system (leading to the so-called first-pass effect), and, as mixing of the venous blood with arterial blood from the liver occurs, consideration and caution are called for in estimating the amounts of chemical in both the systemic circulation and the liver itself.

Despite the GI absorption characteristics discussed earlier, it is common for absorption from the alimentary tract to be facilitated by dilution of the toxicant. Borowitz *et al.* (1971) have suggested that the concentration effects they observed in atropine sulfate, aminopyrine, sodium salicylate, and sodium pentobarbital were due to a combination of rapid stomach emptying and the large surface area for absorption of the drugs.

Major structural or physiological differences in the alimentary tract (e.g., species differences or surgical effects) can give rise to modifications of toxicity. For example, ruminant animals may metabolize toxicants in the GI tract in a way that is unlikely to occur in nonruminants.

The presence of bile salts in the alimentary tract can affect absorption of potential toxicants in a variety of ways, depending on their solubility characteristics.

3.2.3.3 Factors Affecting Absorption

Oral is certainly the most convenient route, and it is the only one of practical importance for self-administration. Absorption, in general, takes place along the whole length of the GI tract, but the chemical properties of each molecule determine whether it will be absorbed in the strongly acidic stomach or in the nearly neutral intestine. Gastric absorption is favored by an empty stomach, in which the chemical, in undiluted gastric juice, will have good access to the mucosal wall. Only when a chemical would be irritating to the gastric mucosa it is rational to administer it with or after a meal. However, the antibiotic griseofulvin is an example of a substance with poor water solubility, the absorption of which is aided by a fatty meal. The large surface area of the intestinal villi, the presence of bile, and the rich blood supply all favor intestinal absorption of griseofulvin and physicochemically similar compounds (Hogben *et al.*, 1959).

The presence of food can impair the absorption of chemicals given by mouth. Suggested mechanisms include reduced mixing, complexing with substances in the food, and retarded gastric emptying. In experiments with rats, prolonged fasting has been shown to diminish the absorption of several chemicals, possibly by deleterious effects upon the epithelium of intestinal villi.

The principles governing the absorption of drugs from the GI lumen are the same as for the passage of drugs across biological membranes elsewhere. Low degree of ionization, high lipid–water partition coefficient of the nonionized form, and small atomic or molecular radii of water-soluble substances all favor rapid absorption. Water passes readily in both directions across the wall of the GI lumen. Sodium ion is probably transported actively from the lumen into the blood. Magnesium ion is very poorly absorbed and therefore acts as a cathartic, retaining an osmotic equivalent of water as it passes down the intestinal tract. Ionic iron is absorbed as an amino acid complex, at a rate usually determined by the body's need for it. Glucose and amino acids are transported across the intestinal wall by specific carrier systems. Some compounds of high molecular weight (polysaccharides and large proteins) cannot be absorbed until they are degraded enzymatically. Other substances cannot be absorbed because they are destroyed by GI enzymes—insulin, epinephrine, and histamine are examples. Substances that form insoluble precipitates in the GI lumen or that are insoluble either in water or in lipid clearly cannot be absorbed.

3.2.3.3.1 Absorption of Weak Acids and Bases

Human gastric juice is very acidic (about pH 1), whereas the intestinal contents are nearly neutral (actually very slightly acidic). The pH difference between plasma (pH 7.4) and the lumen of the GI tract plays a major role in determining whether a drug that is a weak electrolyte will be absorbed into plasma or excreted from plasma into the stomach or intestine. For practical purposes, the mucosal lining of the GI tract is impermeable to the ionized form of a weak acid or base, but the nonionized form equilibrates freely. The rate of equilibration of the nonionized molecule is directly related to its lipid solubility. If there is a pH difference across the membrane, then the fraction ionized may be considerably greater on one side than on the other. At equilibrium, the concentration of the nonionized moiety will be the same on both sides, but there will be more total drug on the side where the degree of ionization is greater. This mechanism is known as *ion trapping*. The energy for sustaining the unequal chemical potential of the acid or base in question is derived from whatever mechanism maintains the pH difference. In the stomach, this mechanism is the energy-dependent secretion of hydrogen ions.

Consider how a weak electrolyte is distributed across the gastric mucosa between plasma (pH 7.4) and gastric fluid (pH 1.0). In each compartment, the Henderson–Hasselbalch equation gives the ratio of acid–base concentrations. The negative logarithm of the acid dissociation constant is designated here by the symbol pK_a rather than the more precisely correct pK^1:

$$pH = pK_a + \log\frac{(\text{base})}{(\text{acid})}$$

$$\log\frac{(\text{base})}{(\text{acid})} = pH - pK_a$$

$$\frac{(\text{base})}{(\text{acid})} = \text{antilog}(pH - pK_a)$$

The implications of the aforementioned equations are clear. Weak acids are readily absorbed from the stomach. Weak bases are not absorbed well; indeed, they would tend to accumulate within the stomach at the expense of agent in the bloodstream. Naturally, in the more alkaline intestine, bases would be absorbed better, and acids more poorly.

It should be realized that although the principle outlines here are correct, the system is dynamic, not static. Molecules that are absorbed across the gastric or intestinal mucosa are removed constantly by blood flow; thus, simple reversible equilibrium across the membrane does not occur until the agent is distributed throughout the body.

Absorption from the stomach, as determined by direct measurements, conforms, in general, to the principles outlined earlier. Organic acids are absorbed well since they are all almost completely nonionized at the gastric pH; indeed, many of these substances are absorbed well since they are all almost completely nonionized at the gastric pH; indeed, many of these substances are absorbed faster than ethyl alcohol, which had long been considered one of the few compounds that were absorbed well from the stomach. Strong acids whose pK_a values lie below 1, which are ionized even in the acid contents of the stomach, are not absorbed well. Weak bases are absorbed only negligibly, but their absorption can be increased by raising the pH of the gastric fluid.

As for bases, only the weakest are absorbed to any appreciable extent at normal gastric pH, but their absorption can be increased substantially by neutralizing the stomach contents. The quaternary cations, however, which are charged at all pH values, are not absorbed at either pH.

The accumulation of weak bases in the stomach by ion trapping mimics a secretory process; if the drug is administered systemically, it accumulates in the stomach. Dogs given various drugs intravenously by continuous infusion to maintain a constant drug level in the plasma had the gastric contents sampled by means of an indwelling catheter. The results showed that stronger bases ($pK_a > 5$) accumulated in stomach contents to many times their plasma concentrations; the weak bases appeared in about equal concentrations in gastric juice and in plasma. Among the acids, only the weakest appeared in detectable amounts in the stomach. One might wonder why the strong bases, which are completely ionized in gastric juice and which theoretical concentration ratios (gastric juice/plasma) are very large, should nevertheless attain only about a 40-fold excess over plasma. Direct measurements of arterial and venous blood show that essentially all the blood flowing through the gastric mucosa is cleared of these agents; obviously, no more chemical can enter the gastric juice in a given time period than is brought there by circulation. Another limitation comes into play when the base pK_a exceeds 7.4; now a major fraction of the circulating base is cationic, and a decreasing fraction is nonionized, so the effective concentration gradient for diffusion across the stomach wall is reduced (Kerberle, 1971).

The ion-trapping mechanism provides a method of some forensic value for detecting the presence of alkaloids (e.g., narcotics, cocaine, amphetamines) in cases of death suspected to be due to overdosage of self-administered drugs. Drug concentrations in gastric contents may be very high even after parenteral injection.

Absorption from the intestine has been studied by perfusing drug solutions slowly through rat intestine *in situ* and by varying the pH as desired. The relationships that emerge from such studies are the same as those for the stomach, the difference being that the intestinal pH is normally very near neutrality. As the pH is increased, the bases are absorbed better, and acids more poorly. Detailed studies with a great many drugs in unbuffered solutions revealed that in the normal intestine, acids with $pK_a > 3.0$ and bases with $pK_a < 7.8$ are very well absorbed; outside these limits the absorption of acids and bases falls off rapidly. This behavior leads to the conclusion that the "virtual pH" in the microenvironment of the absorbing surface in the gut is about 5.3; this is somewhat more acidic than the pH in the intestinal lumen is usually considered to be.

Absorption from the buccal cavity has been shown to follow exactly the same principles as those described for absorption from the stomach and intestine.

The pH of human and canine saliva is usually about 6. Bases in people are absorbed only on the alkaline side of their pK_a, that is, only in the nonionized form. At normal saliva pH, only weak bases are absorbed to a significant extent.

3.2.4 Bioavailability and Thresholds

The difference between the extent of availability (often designated solely as bioavailability) and the rate of availability is illustrated in Figure 3.4, which depicts the concentration–time curve for a hypothetical agent formulated into three different dosage forms. Dosage forms A and B are designed so that the agent is put into the blood circulation at the same rate but twice as fast as for dosage form C. The times at which agent concentrations reach a peak are identical for dosage forms A and B and occur earlier than the peak time for dosage form C. In general, the relative order of peak times following the administration of different dosage forms of the drug corresponds to the rates of availability of the chemical moiety from the various dosage forms. The extent of availability can be measured by using either chemical concentrations in the plasma or blood or amounts of unchanged chemical in the urine. The area under the blood concentration–time curve for an agent can serve as a measure of the extent of its availability. In Figure 3.4, the areas under curves A and C are identical and twice as great as the area under curve B. In most cases, where clearance is constant, the relative areas under the curves or the amount of unchanged chemical excreted in the urine will quantitatively describe the relative availability of the agent from the different dosage forms. However, even in nonlinear cases, where clearance is dose dependent, the relative areas under the curves will yield a measurement of the rank order of availability from different dosage forms or from different routes of administration.

Because there is usually a critical concentration of a chemical in the blood that is necessary to elicit either a pharmacological or toxic effect, both the rate

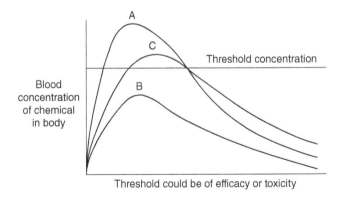

Figure 3.4 Three different systemic absorption curves.

and extent of input or availability can alter the toxicity of a compound. In the majority of cases, the duration of effects will be a function of the length of time the blood concentration curve is above the threshold concentration; the intensity of the effect for many agents will be a function of the elevation of the blood concentration curve above the threshold concentration.

Thus, the three different dosage forms depicted in Figure 3.4 will exhibit significant differences in their levels of "toxicity." Dosage form B requires that twice the dose be administered to attain blood levels equivalent to those for dosage form A. Differences in the rate of availability are particularly important for agents given acutely. Dosage form A reaches the target concentration earlier than chemical from dosage form C; concentrations from A reach a higher level and remain above the minimum effect concentration for a longer period of time. In a multiple dosing regimen, dosage forms A and C will yield the same average blood concentrations, although dosage form A will show somewhat greater maximum and lower minimum concentrations.

For most chemicals, the rate of disposition or loss from the biological system is independent of rate of input, once the agent is absorbed. Disposition is defined as what happens to the active molecule after it reaches a site in the blood circulation where concentration measurements can be made (the systemic circulations, generally). Although disposition processes may be independent of input, the inverse is not necessarily true, because disposition can markedly affect the extent of availability. Agents absorbed from the stomach and the intestine must first pass through the liver before reaching the general circulation (Figure 3.3). Thus, if a compound is metabolized in the liver or excreted in bile, some of the active molecule absorbed from the GI tract will be inactivated by hepatic processes before it can reach the systemic circulation and be distributed to its sites of action. If the metabolizing or biliary-excreting capacity of the liver is great, the effect on the extent of availability will be substantial. Thus, if the hepatic blood clearance for the chemical is large, relative to hepatic blood flow, the extent of availability for this chemical will be low when it is given by a route that yields first-pass metabolic effects. This decrease in availability is a function of the physiological site from which absorption takes place, and no amount of modification to dosage form can improve the availability under linear conditions. Of course, toxic blood levels can be reached by this route of administration if larger doses are given.

It is important to realize that chemicals with high extraction ratios (i.e., greater extents of removal by the liver during first-pass metabolism) will exhibit marked intersubject variability in bioavailability because of variations in hepatic function or blood flow or both. For the chemical with an extraction ratio of 0.90 that increases to 0.95, the bioavailability of the agent will be halved, from 0.10 to 0.05. These relationships can explain the marked variability in plasma or blood drug concentrations that occur among individual animals given similar doses of a chemical that is markedly extracted. Small variations in hepatic extraction between individual animals will result in large differences in availability and plasma drug concentrations.

The first-pass effect can be avoided, to a great extent, by use of the sublingual route and by topical preparations (e.g., nitroglycerine ointment), and it can be partially avoided by using rectal suppositories. The capillaries in the lower and middle sections of the rectum drain into the interior and middle hemorrhoidal veins, which in turn drain into the inferior vena cava, thus bypassing the liver. However, suppositories tend to move upward in the rectum into a region where veins that lead to the liver predominate, such as the superior hemorrhoidal vein. In addition, there are extensive connections between the superior and middle hemorrhoidal veins, and thus probably only about 50% of a rectal dose can be assumed to bypass the liver. The lungs represent a good temporary clearing site for a number of chemicals (especially basic compounds) by partition into lipid tissues, as well as serve a filtering function for particulate matter that may be given by intravenous injection. In essence, the lung may cause first-pass loss by excretion and possible metabolism for chemicals input into the body by the non-GI routes of administration.

Biological (test subject) factors that can influence absorption of a chemical from the GI tract are summarized in Table 3.4.

There are also a number of chemical factors that may influence absorption from the GI tract. These are summarized in Table 3.5.

3.3 Formulation Supporting Toxicology Studies

Formulations used to administer potential drugs undergoing development occupy an unusual place in pharmaceutical safety assessment compared to the rest of the industrial toxicology. Eventually, a separate function in the pharmaceutical company developing a drug will develop a specific formulation that is to be administered to people—a formulation that optimizes the conditions of absorption and stability for the drug entity (Racy, 1989). The final formulation will need to be assessed to see if it presents any unique local or short-term hazards, but as long as its nonactive constituents are drawn from the approved formulary lists, no significant separate evaluation of their safety is required preclinically. They can, of course, alter the toxicity of the drug under study.

Simultaneous with this development of an optimized clinical formulation, however, preclinical evaluations of the safety of the drug moiety must be performed and both have slightly different requirements (multiples of what will be clinical doses must be delivered) and have a wider range of potential vehicles to use (i.e., some such as DMSO, which cannot be used in humans or for which such use is extremely limited by regulations). Separate preclinical formulations (which generally are less complex than the clinical ones) are developed,

Table 3.4 Selected factors that may affect chemical distribution to various tissues.

A) Factors relating to the chemical and its administration
  1) Degree of binding of chemical to plasma proteins (i.e., agent affinity for proteins) and tissues
  2) Chelation to calcium, which is deposited in growing bones and teeth (e.g., tetracyclines in young children)
  3) Whether the chemical distributes evenly throughout the body (one-compartment model) or differentially between different compartments (models of two or more compartments)
  4) Ability of chemical to cross the blood–brain barrier
  5) Diffusion of chemical into the tissues or organs and degree of binding to receptors that are and are not responsible for the drug's beneficial effects
  6) Quantity of chemical given
  7) Route of administration or exposure
  8) Partition coefficients (nonpolar chemicals are distributed more readily to fat tissues than are polar chemicals)
  9) Interactions with other chemicals that may occupy receptors and prevent the drug from attaching to the receptor, inhibit active transport, or otherwise interfere with a drug's activity
  10) Molecular weight of the chemical
  11) Is there a transporter for drug structural class?
B) Factors relating to the test subject
  1) Body size
  2) Fat content (e.g., obesity affects the distribution of drugs that are highly soluble in fats)
  3) Permeability of membranes
  4) Active transport for chemicals carried across cell membranes by active processes
  5) Amount of proteins in blood, especially albumin
  6) Pathology or altered homeostasis that affects any of the other factors (e.g., cardiac failure, renal failure)
  7) Presence of competitive binding substances (e.g., specific receptor sites in tissues bind drugs)
  8) pH of blood and body tissues
  9) pH of urine[a]
  10) Blood flow to various tissues or organs (e.g., well-perfused organs usually tend to accumulate more chemical than less well-perfused organs)

[a] The pH of urine is usually more important than the pH of blood.

sometimes by a formulation group and other times by the toxicology group itself. These preclinical formulations will frequently include much higher concentrations of the drug moiety being tested than do any clinical formulations. The preclinical formulations are developed and evaluated with the aim of reproducibly delivering the drug (if at all possible by the route intended in man).

Table 3.5 Chemical characteristics of a drug that may influence absorption.

A) Administration of chemical and its passage through the body

1) Dissolution characteristics of solid dosage forms, which depend on formulation in addition to the properties of the chemical itself (e.g., vehicle may decrease permeability of suspension or capsule to water and retard dissolution and diffusion)

2) Rate of dissolution in gastrointestinal fluids. Chemicals that are inadequately dissolved in gastric contents may be inadequately absorbed

3) Chemicals that are absorbed into food may have a delayed absorption

4) Carrier-transported chemicals are more likely to be absorbed in the small intestine

5) Route of administration

6) Chemicals undergo metabolism in the gastrointestinal tract

B) Physiochemical properties of chemicals

1) Chemicals that chelate metal ions in food may form insoluble complexes and will not be adequately absorbed

2) pH of dosing solutions—weakly basic solutions are absorbed to a greater degree in the small intestine

3) Salts used

4) Hydrates or solvates

5) Crystal form of chemical (e.g., insulin)

6) "Pharmaceutical" form (e.g., fluid, solid, suspension)

7) Enteric coating

8) Absorption of quaternary compounds (e.g., hexamethonium, amiloride) is decreased by food

9) Molecular weight of chemical (e.g., when the molecular weight of a drug is above about 1000, absorption is markedly decreased)

10) pK_a (dissociation constant)

11) Lipid solubility (i.e., a hydrophobic property relating to penetration through membranes)

12) Particle size of chemical in solid dosage form—smaller particle sizes will increase the rate and/or degree of absorption if dissolution of the chemical is the rate-limiting factor in absorption. Chemicals that have a low dissolution rate may be made in a micronized form to increase their rate of dissolution

13) Particle size of the dispersed phase in an emulsion

14) Type of disintegrating agent in the formulation

15) Hardness of a solid (granule, pellet, or tablet) (i.e., related to amount of compression used to make tablet) or capsule if they do not disintegrate appropriately

Maintaining drug stability through an optimum period of time and occluding the observed effects of the drug with vehicle effects to the minimum extent possible should be the objective of the formulation. And these preclinical formulations are not restricted to materials that will (or even can) be used in final clinical formulations (Boersen *et al.*, 2014; Gad, 2008a).

In the GLP general toxicology studies (sometimes called pivotal toxicity studies), the actual blood levels of active moiety that are achieved will be determined so that correlations to later clinical studies can be made. It should be noted that such plasma levels are a surrogate for what we are really interested in—levels at target organ or tissue sites. But plasma is convenient to collect without compromising the test animals.

The formulations that are developed and used for preclinical studies are sometimes specific for the test species to be employed, but their development always starts with consideration of the route of exposure that is to be used clinically and, if possible, in accordance with a specified regimen of treatment (mirroring the intended clinical protocol as much as possible). One aspect of both nonclinical and clinical formulation and testing that prevents an important but often overlooked aspect of pharmaceutical safety assessment is the special field of excipients. These will be considered at the end of this chapter.

Among the cardinal principles of both toxicology and pharmacology is that the means by which an agent comes in contact with or enters the body (i.e., the route of exposure or administration and the regimen of administration) does much to determine the nature and magnitude of an effect. However a rigorous understanding of formulations, routes, and their implications in the design and analysis of safety studies is not widespread. And in the day-to-day operations of performing studies in animals, such an understanding of routes, their manipulation, means and pitfalls of achieving them, and art and science of vehicles and formulations is essential to the sound and efficient conduct of a study.

Many use the Biopharmaceutics Classification System (BCS) from the FDA Office of Pharmaceutical Sciences (USFDA, 2005) with its four classifications (classes 1–4) of oral molecules (based on solubility and intestinal permeability) as a starting tool to understand potential development risks due to its potential to predict intestinal drug absorption. Class 1 (high solubility, high permeability) and class 2 (low/moderate solubility, high permeability) present the easiest path to achieving systemic absorption via the oral route, and therefore those molecules which meet these criterion should be favored in selection for development over those that don't. Of course, these are but two factors to be considered in such selections (Gad, 2005; Liu, 2008).

There are multiple computational methods to predict potential product characteristics based on structural aspects particularly physicochemical aspects such as solubility, pK_a, pH, and intestinal permeability (Selassie and Verma, 2010). Using these tools can serve as part of the tool set to optimize the selection of lead candidates into development.

3.3.1 Formulation of Test Materials

As an integral part of the formulation development in the early stage to candidate selection, it is important to make sure that the formulation used for toxicological studies is considered as part of the overall set of preclinical animal studies. To ensure comparability with pharmacokinetic and pharmaco-dynamic studies, the following considerations are needed:

1) The formulation has been characterized both chemically and physically.
2) Sufficient exposure in the relevant species had been demonstrated including administration after multiple dosing.
3) Dose linearity not necessarily proportional at least for lower dose levels has also been shown.
4) Ease of administration is assessed that this will not be detrimental to the repeated dosing in the species selected.

Once we quantitate or estimate the necessary approximate human therapeutic dose, then the concept of the maximum absorbable dose (MAD) allows us to identify what the "carrying concentration" of a formulation must be (Strickley, 2008). That is what quantity of drug per milliliter of formulation must be achieved to meet the needs of delivering enough drugs into the systemic circulation.

One of the areas that is overlooked by virtually everyone in toxicology testing and research, yet is of crucial importance, is the need for formulation of candidate drugs and the use of vehicles and excipients in the formulation of test chemicals for administration to test animals (Strickley, 2008). For a number of reasons, a drug of interest is rarely administered or applied as is ("neat"). Rather, it must be put in a form that can be accurately given to animals in such a way that it will be absorbed and not be too irritating. Most laboratory toxicologists come to understand vehicles and formulation, but to the knowledge of the author, guidance on the subject is limited to a short chapter on formulations by Fitzgerald *et al.* (1983). There is also a very helpful text on veterinary dosage forms by Hardee and Baggo (1998). Available vehicles and their tolerated use doses and concentrations are presented in codex articles (Gad, 2015; Gad *et al.*, 2006; Thackaberry, 2013). One approach is to reduce the size of solid drug particulates as much as possible—most recently nanosuspensions have become popular (Rabinow, 2004).

More advanced formulation development options from pharmaceutical industries for brick dust solubility compounds emerged lately in literature (Hageman, 2010; Higgins *et al.*, 2012; Maas *et al.*, 2007; Neervannan, 2006; Palucki *et al.*, 2010; Thackaberry, 2013). The reviews cover the interrelationship of physicochemical properties such as solubility, partition coefficient, and permeability to the performance of the API and preclinical studies. Formulation development strategies take into consideration the maximum formulatable dose, BCS of the drug, intended duration of action, multiple dosing, and desired route of administration. Commonly used solubilizers, surfactants, cosolvents, and acceptable pH range of the formulations and adherence to regulatory requirements are presented. Thackaberry (2013) reviewed oral preclinical vehicles based on data from the FDA's Orange Book with commonly used vehicles used in repeat dose oral toxicity. Palucki *et al.* (2010) provided Merck's perspective of using decision tree in guiding formulation selection of poorly soluble compounds. Another systematic approach for formulation selection was presented by Lohani *et al.* (2014), using GPR119 agonist, a BCS II compound as an example of a challenging compound. They described the approach after identifying a new non-bioavailable polymorph (form B) to improving the exposure using an enabling formulation such as amorphous solid dispersion. Most of the references described the continuum of the pre-clinical formulation from a single dose evaluation to a repeat dosing of the preclinical studies. Opportunity to improve bioavailability by use of amorphous form had gained momentum recently. However since amorphous form is highly metastable state where there is a thermodynamic drive toward crystallization, preparation of dispersion of the drug in pharmaceutically acceptable polymers such as HPMC and PVP is available to stabilize against crystallization (Le-Ngoc Vo *et al.*, 2013; Newman *et al.*, 2012; Sotthivirat *et al.*, 2013). With the advent of spray-drying technologies, quantities of a few grams to tens of kilograms are possible. Spray drying is one of the most common techniques used to prepare solid dispersions due to possibility of continuous manufacturing, ease of scalability, good uniformity of molecular dispersion, and high recoveries. Consequently, these spray-dried dispersions can be used from preclinical studies to clinical as intermediates to be further processed into conventional tablets. Commercial products such as Kaletra®, Prograf®, and Cesamet® are based on solid dispersions.

Table 3.6 presents an overview of typical forms of excipients for oral dosage forms. The entire process of drug development—even during the preclinical and nonclinical phase—includes a continuous development (with increased sophistication) of formulation. The use of nanoparticles (actually, as seeking even finer *micronized* particles in dosing formulations, it has been around for decades) is the latest approach.

Regulatory toxicology in the United States can be said to have arisen, due to the problem of vehicles and formulation, in the late 1930s, when attempts were made to formulate the new drug sulphanilamide. This drug is not very soluble in water, and a US firm called Massengill produced a clear, syrupy elixir formulation that was easy to take orally. The figures illustrate how easy it is to be misled. The drug sulphanilamide is not very soluble in glycerol, which has an LD_{50} in mice of 31.5 g/kg, but there are other glycols that have the characteristic sweet taste and a much higher solvent capacity. Ethylene glycol has an LD_{50} of 13.7 g/kg in mice and 8.5 g/kg in rats, making it slightly more toxic than diethylene glycol, which has an LD_{50} in rats of 20.8 g/kg, similar to that for glycerol. The drug, which is itself inherently toxic, was marketed in a 75% aqueous diethylene glycol-flavored elixir. Early in 1937 came the first reports of deaths, but the situation remained obscure for about 6 months until it became clear that the toxic ingredient in the elixir was the diethylene glycol. Even as late as March 1937, Haag and Ambrose were reporting that the glycol was excreted substantially unchanged in dogs, suggesting that it was likely to be

Table 3.6 Non-active formulation components in drug candidate oral delivery.

Type	Purpose	Examples
	Tablets, sachets, and capsules	
Binder	Facilitates agglomeration of powder into granules	Povidones, starches
Capsule shell	Contains powders or liquids	Gelatin, hypromellose
Coating agent	May mask unpleasant tastes or odors, improve ingestion or appearance, protect ingredients from the environment, or modify release of the active ingredient	Shellac, hypromellose
Colorant	Produces a distinctive appearance and may protect light-sensitive ingredients	FD&C colors, titanium dioxide
Disintegrant	Promotes rapid disintegration to allow a drug to dissolve faster	Sodium starch glycolate, crospovidones
Enteric coatings	Protects from dissolution in stomach ("gastro resistant")	HPMC, methylacrylate copolymers
Filler or diluent	Increases volume or weight	Calcium phosphate, lactose
Glidant or anticaking agent	Promotes powder flow and reduces caking or clumping	Talc, colloidal silicon dioxide
Lubricant	Reduces friction between particles themselves and between particles and manufacturing equipment	Magnesium stearate, glycerides
Release modifier	Provides extended-release capability	Ethylcellulose, guar gum
Solvent/vehicle	Improves stability and bioavailability	Water++ (see Gad, 2016)
Surfactant	Solubilizing agent	Irganox, SLS
Oral liquids		
Antimicrobial preservative	Prevents growth of bacteria, yeast, and mold	Glycerin, benzyl alcohol
Antioxidant	Reduces oxidative reactions that could alter ingredients	Ascorbic acid, butylated hydroxyanisole
Chelating or complexing agent	Stabilizes ions	Ethylenediaminetetraacetic acid salts, cyclodextrins
Liposome preparations	Improve bioavailability and protect from first pass metabolism	Special vehicle with at least one lipid bilayer
pH modifier	Controls pH to improve drug stability or avoid irritation when consumed	Citric acid and its salts, salts of phosphoric acid
Surfactants and solubilizing agent	Promotes dissolution of insoluble ingredients	Sodium lauryl sulfate, polysorbates
Sweetening agent	Improves palatability	Sucrose, saccharin

Source: Adapted from Gad (2008b, 2016). Reproduced with permission of John Wiley & Sons.

safe (Hagenbusch, 1937). Within a few weeks, Holick (1937) confirmed that a low concentration of diethylene in drinking water was fatal to a number of species. Hagenbusch (1937) found that the results of necropsies performed on patients who had been taking 60–70 mL of the solvent per day were similar to those of rats, rabbits, and dogs taking the same dose of solvent with or without the drug. This clearly implicated the solvent, although some authors considered that the solvent was simply potentiating the toxicity of the drug. Some idea of the magnitude of this disaster may be found in the paper of Calvary and Klump (1939), who reviewed 105 deaths and a further 2560 survivors who were affected to varying degrees, usually with progressive failure of the renal system. It is easy to be wise after the event, but the formulator fell into a classic trap, in that the difference between acute and chronic toxicity had not been adequately considered. In passing, the widespread use of ethylene glycol itself as an antifreeze has led to a number of accidental deaths, which suggests that the lethal dose in man is around 1.4 mL/kg or a volume of about 100 mL. In the preface to the first United States Pharmacopeia (USP), published in 1820, there is a the statement that "It is the object of the Pharmacopoeia to select from among substances which possess medical power, those, the utility of which is most fully established and best understood; and to form from them preparations and compositions, in which their powers may be exerted to the greatest advantage." This statement suggests that the influence that formulation and preparation may have on the biological activity of a drug (and on nonpharmaceutical chemicals) has been appreciated for a considerable time.

There are some basic principles to be observed in developing and preparing test material formulations. These are presented in Table 3.7. A start to all of this should be preformulation—characterizing the chemical, physical, and physical chemistry aspects of the drug molecule (Neervannan, 2006). Parts of this are identifying the optimal phase of the drug for formulation (Palucki *et al.*, 2010) and useful salts (Stahl and Wermuth, 2011; Thackaberry, 2012). Oral formulation, which is usually a dosage form suspended in aqueous medium, is manually prepared by conventional method using a mortar and pestle during the early small preparation scale pharmacokinetic studies when restricted amounts of compounds are available. As the discovery phase progressed to the preclinical toxicology studies, the highest doses are increased to typically 100-fold of ED_{50} or to the FDA recommended maximum of 2 g/kg where the compound does not exhibit adverse effects at preclinical safety studies (Neervannan, 2006). In addition the dosing period is prolonged up to 28 days and that a non-rodent species such as dogs or monkeys are also included in the studies. Such situation significantly increases the preparation scale to 2 L of suspension. Mixing is a critical quality to be maintained with consideration to the compound morphology, polymorph, and batch to batch differences, which will likely cause heterogeneous drug distribution and hence variable dosing. While early discovery preparation can be assessed visually, larger-scale

Table 3.7 Basic principles to be observed in developing and preparing test material formulations.

A) Preparation of the formulation should not involve heating of the test material anywhere near to the point where its chemical or physical characteristics are altered

B) If the material is a solid and it is to be assessed for dermal effects, its shape and particle size should be preserved. If intended for use in man, topical studies should be conducted with the closest possible formulation to that to be used on humans

C) Multicomponent test materials (mixtures) should be formulated so that the administered form accurately represents the original mixture (i.e., components should not be selectively suspended or taken into solution)

D) Formulation should preserve the chemical stability and identity of the test material

E) The formulation should be such as to minimize total test volumes. Use just enough solvent or vehicle

F) The formulation should be easy to administer accurately

G) pH of dosing formulations should be between 5 and 9, if possible

H) Acids or bases should not be used to divide the test material (for both humane reasons and to avoid pH partitioning in either the gut or the renal tubule)

I) If a parental route is to be employed, final solutions should be as nearly isotonic as possible. Do not assume a solution will remain such upon injection into the blood stream. It is usually a good idea to verify that the drug stays in solution upon injection by placing some drops into plasma

J) Particularly if use is to be more than a single injection, steps (such as filtration) should be taken to ensure suitable sterility

preparation needed more sophisticated testing and should be included in the toxicology protocol. Besides homogeneity assessment, chemical and physical stability and particle size distribution of the suspension should also be tested.

3.3.2 Practices in Oral Preclinical Formulations

The physical form of a material destined for oral administration often presents unique challenges. Liquids can be administered as supplied or diluted with an appropriate vehicle, and powders or particulates can often be dissolved or suspended in an appropriate vehicle. However, selection of an appropriate vehicle is often difficult. Water and oil (such as the vegetable oils) are used most commonly. Materials that are not readily soluble in either water or oil can frequently be suspended in a 1% aqueous mixture of methylcellulose. Occasionally, a more concentrated methylcellulose suspension (up to 5%) may be necessary. Materials for which appropriate solutions or suspensions cannot be prepared using one of these three vehicles often present major difficulties.

Limited solubility or suspendability of a material often dictates preparation of dilute mixtures that may require large volumes to be administered. The total volume of liquid dosing solution or suspension that can be administered to a rodent is limited by the size of its stomach. However, because rats lack a gagging reflex and have no emetic mechanism, any material administered will be retained. Guidelines for allowable excipients to use for oral formulation are given in Table 3.6.

Limitations on total volume, therefore, present difficulties for materials that cannot easily be dissolved or suspended. The most dilute solutions that can be administered for a limit-type test (5000 mg/kg), using the maximum volumes shown in Table 3.8, generally are 1% for aqueous mixtures and 50% for other vehicles.

Although "vehicle control" animals are not required for commonly used vehicles (water, oil, methylcellulose), most regulations require that the biological properties of a vehicle be known and/or that historical data be available. Unfortunately, the best solvents are generally toxic and, thus, cannot be used as vehicles. Ethanol and acetone can be tolerated in relatively high doses but produce effects that may complicate interpretation of toxicity associated with the test material alone. It is sometimes possible to dissolve a material in a small amount of one of these vehicles and then dilute the solution in water or in oil.

Other possibilities for insoluble materials are to mix the desired amount of material with a small amount of the animal's diet or to use capsules. The difficulty with the diet approach is the likelihood that the animal will not consume all of the treated diet or that it may selectively not consume chunks of test material. Use of capsules, meanwhile, is labor intensive. In rare cases, if all of these approaches fail, it may not be possible to test a material by oral administration. In capsules, particle size is generally inversely related to solubility and

Table 3.8 General guidelines for maximum dose volumes by route.

Route	Volume (mL/kg) should not exceed	Notes
Oral	20 (at one time)	Fasted animals
Dermal	2	Limit is accuracy of dosing per available body surface
Intravenous	1	Over 5 min
Intramuscular	0.5	At one site
Periocular	0.01 mL	
Rectal	0.5	
Vaginal	0.2 mL in rat	
	1 mL in rabbit	
Inhalation	2 mg/L	
Nasal	0.1 mL/nostril in monkey or dog	

Source: Baker *et al.* (1979). Reproduced with permission of Springer.

bioavailability. However, milling of solids may adversely affect their chemical nature and/or pose issues of safety.

If necessary, the test substance should be dissolved or suspended as a suitable vehicle, preferably in water, saline, or an aqueous suspension such as 0.5% methyl cellulose in water. If a test substance cannot be dissolved or suspended in an aqueous medium to form a homogenous dosage preparation, corn oil or another solvent can be used. The animals in the vehicle control group should receive the same volume of vehicle given to animals in the highest dose group.

The test substance can be administered to animals at a constant concentration across all dose levels (i.e., varying the dose volume) or at a constant dose volume (i.e., varying the dose concentration). However, the investigator should be aware that the toxicity observed by administration in a constant concentration may be different from that observed when given in a constant dose volume. For instance, when a large volume of corn oil is given orally, GI motility is increased, causing diarrhea and decreasing the time available for absorption of the test substance in the GI tract. This situation is particularly true when a highly lipid-soluble chemical is tested.

If an organic solvent is used to dissolve the chemical, water should be added to reduce the dehydrating effect of the solvent within the gut lumen. The volume of water or solvent–water mixture used to dissolve the chemical should be kept low, since excess quantities may distend the stomach and cause rapid gastric emptying. In addition, large volumes of water may carry the chemical through membrane pores and increase the rate of absorption. Thus, if dose-dependent absorption is suspected, it is important that the different doses are given in the same volume of solution.

Large volumes than those detailed earlier may be given, although nonlinear kinetics seen under such circumstances may be due to solvent-induced alteration of intestinal function. The use of water-immiscible solvents such as corn oil (which are sometimes used for gavage doses) should be avoided, since it is possible that mobilization from the vehicle may be rate limiting. Magnetic stirring bars or homogenizers can be used in preparing suspensions. Sometimes a small amount of a surfactant such as Tween 80, Span 20, or Span 60 is helpful in obtaining a homogenous suspension.

A large fraction of such a material may quickly pass through the GI tract and remain unabsorbed. Local irritation by a test substance generally decreases when the material is diluted. If the objective of the study is to establish systemic toxicity, the test substance should be administered in a constant volume to minimize GI irritation that may, in turn, affect its absorption. If, however, the objective is to assess the irritation potential of the test substance, then it should be administered undiluted.

As pharmaceutical excipients are assumed to be biologically nonreactive, dose–response relations cannot always be established. An acceptable alternative is to use a maximum attainable or maximum feasible dose. This is the highest dose possible that will not compromise the nutritional or health status of the animal. Table 3.9 summarizes the maximum or limit doses for various types of studies by different routes of exposure. For example, 2000 mg/kg body weight of an orally administered test material is the maximum dose recommended for a testing strategy that has been developed for new pharmaceutical excipients that takes into consideration the physical–chemical nature of the product and the potential route(s) and duration of exposures, both through its intended use as part of a drug products and through workplace exposure during manufacturing. The number and types of studies recommended in this tiered approach are based on the duration and routes of potential human exposure. Thus, the longer the exposure to the new pharmaceutical excipient, the more studies are necessary to assure safety.

Table 3.9 Limit doses for toxicological studies.

Nature of test	Species	Limit dose[a]
Acute oral	Rodent	2000 mg/kg bw
Acute dermal	Rabbit	2000 mg/kg bw
	Rat	
Acute inhalation[b]	Rat	5 mg/L air for 4 h or maximum attainable level under conditions of study
Dermal irritation	Rabbit	0.5 mL liquid
		0.5 g solid
Eye irritation	Rabbit	0.1 mL liquid
		100 mg solid
14-/28-day oral repeated dosing; 90-day subchronic	Rodent, non-rodent	1000 mg/kg bw/day
14-/28-day oral repeated dosing; 90-day subchronic	Rat, rabbit	1000 mg/kg bw/day
Chronic toxicity, carcinogenicity	Rats, mice	5% maximum dietary concentration for nonnutrients
Reproduction	Rats	1000 mg/kg bw/day
Developmental toxicity (teratology)	Mice, rats, rabbits	1000 mg/kg bw/day

Source: Wiener and Katkoskie (1999). Reproduced with permission of Taylor & Francis.
[a] Milligrams of test material dosed per kilogram of body weight to the test species (mg/kg bw).
[b] Acute inhalation guidelines that indicate this limit dose are US Environmental Protection Agency Toxic Substance Health Effect Test Guidelines, October, 1984; (PB82-232984) Acute Inhalation Toxicity Study; the OECD Guidelines of the Testing of Chemicals, Vol. 2, Section 4; Health Effects, 403, Acute Inhalation Toxicity Study, May 12, 1982, and the Official Journal of the European Communities, L383A, Vol. 35, December 29, 1992, Part B.2.

Regulatory guidelines and standard practices identify studies that have been outlined for each exposure category to assure safe use of the time period designated. The tests for each exposure category assure the safe use of the new pharmaceutical excipient of the time frame specified for the specific exposure category. Additional tests are required for longer exposure times.

3.3.3 Excipients/Vehicles

Excipients are usually thought of as inert substances (such as gum arabic and starch) that form the vehicle or bulk of the dosage form of a drug. They are, of course, both much more complicated than this and not necessarily inert. A better definition would be that of the USP and National Formulary (USP-NF, 2014), which defined excipients as any component, other than the active substances (i.e., drug substances or DS) intentionally added to the formulation of a dosage form. These substances serve a wide variety of purposes: enhancing stability, adding bulking, increasing and/or controlling absorption, providing or masking flavor, coloring, and serving as a lubricant in the manufacturing process. They are, in fact, essential for the production and delivery of marketed drug products. As will soon be made clear, they are regulated both directly and as part of the drug product. For the pharmaceutical manufacturers, using established and accepted excipients (such as can be found in Smolinske (1992) or APA (1994)—though these lists are not complete) is much preferred. However, both pharmaceutical manufacturers and the companies that supply excipients must from time to time utilize (and therefore develop, evaluate for safety, and get approved) new excipients.

While for the purpose of nonclinical formulation, our concerns are generally limited to vehicles, their formulation components can be important. Table 3.6 lists examples of this.

3.3.4 Regulation of Excipients

Table 3.10 lists the relevant sections of CFR 21 that governs excipients. Under Section 201(g)(1) of the Federal Food, Drug, and Cosmetic Act (FD&C Act; 1), the term *drug* is defined as

A) Articles recognized in the official *USP*, official *Homeopathic Pharmacopeia of the United States*, or official *National Formulary*, or any supplement to any of them
B) Articles intended for use in the diagnosis, cure, mitigation, treatment, or prevention of disease in man or other animals
C) Articles (other than food) intended to affect the structure of any function of the body of man or other animals
D) Articles intended for use as a component of any articles specified in clause (A), (B), or (C)

Table 3.10 US Code of Federal Register references to excipients.

Subject	Reference	Content
General	21 CFR § 210.3(b)(8)	Definitions
	21 CFR § 201.117	Inactive ingredients
	21 CFR § 210.3(b)(3)	Definitions
Over-the-counter drug products	21 CFR § 330.1(e)	General conditions for general recognition as safe, effective, and not misbranded
	21 CFR § 328	Over-the-counter drug products intended for oral ingestion that contain alcohol
Drug Master Files	21 CFR § 314.420	Drug master files
Investigational New Drug application	21 CFR § 312.23(a)(7)	IND content and format
New drug application	21 CFR § 312.31	Information amendments
	21 CFR § 314.50(d)(1)(ii)(a)	Content and format of an application
	21 CFR § 314.70	Supplements and other changes to an approved application
Abbreviated New Drug Application	21 CFR § 314.94(a)(9)	Content and format of an abbreviated application
	21 CFR § 314.127	Refusal to approve an abbreviated new drug application
	21 CFR § 314.127(a)(8)	Refusal to approve an abbreviated new drug application
Current Good Manufacturing Practice	21 CFR § 211.84(d)	Testing an approval or rejection of components, drug product containers, and closures
	21 CFR § 211.165	Testing and release for distribution
	21 CFR § 211.180(b)	General requirements
	21 CFR § 211.80	General requirements
	21 CFR § 211.137	Expiration dating
Listing of drugs	21 CFR § 207	Registration of procedures of drugs and listing of drugs in commercial distribution
	21 CFR § 207.31(b)	Additional drug listing information
	21 CFR § 207.10(e)	Exceptions for domestic establishments

Table 3.10 (*continued*)

Subject	Reference	Content
Labeling	21 CFR § 201.100(b)(5)	Prescription drugs for human use
	21 CFR § 201.20	Declaration of presence of FD&C Yellow No. 5 and/or FD&C Yellow No. 6 in certain drugs for human use
	21 CFR § 201.21	Declaration of presence of phenylalanine as a component of aspartame in over-the-counter and prescription drugs for human use
	21 CFR § 201.22	Prescription drugs containing sulfites; required warning statements

An excipient meets the definitions as listed in (A) and (D) earlier.

In 21 CFR § 210.3(b)(8)(2), an "inactive ingredient means any component other than an active ingredient." According to the CFR, the term inactive ingredient includes materials in addition to excipients. 21 CFR § 201.117 states the following:

Inactive ingredients: A harmless drug that is ordinarily used as an inactive ingredient such as a coloring, emulsifier, excipient, flavoring, lubricant, preservative, or solvent in the preparation of other drugs shall be exempt from Section 502(f) (1) of the Act. This exemption shall not apply to any substance intended for a use, which results in the preparation of a new drug, unless an approved new drug application provides for such use.

Excipients also meet the definition of a component in the good manufacturing practice (GMP) regulations in 21 CFR § 210.3(b)(3): "Component means any ingredient intended for use in the manufacture of a drug product, including those that may not appear in such drug product."

The *NF* admissions policy in the *USP-NF 25* defines the word *excipient* (3): "An excipient is any component other than the active substance(s), intentionally added to the formulation of a dosage form. It is not defined as an inert commodity or an inert component of a dosage form."

Similar to all other drugs, excipients must comply with the adulteration and misbranding provisions of the FD&C Act. Under Section 501(a), an excipient shall be deemed to be adulterated if it consists in whole or in part of any filthy, putrid, or decomposed substance or if it has been prepared, packed, or held under unsanitary conditions whereby it may have been contaminated with filth or whereby it may have been rendered injurious to health. An excipient is adulterated if the methods used in or the facilities or controls used for its manufacture, processing, packing, or holding do not conform to or are not operated or administered in conformity with current GMP to assure that such drug meets the requirements of the act as to safety and has the identity and strength and meets the quality and purity characteristics, which it purports or is represented to possess. In addition, under Section 501(b), an excipient shall be deemed to be adulterated if it purports to be or is represented as a drug, the name of which is recognized in an official compendium and its strength differs from, or its quality or purity falls below, the standards set forth in such compendium (Katdare and Chaubal, 2006).

In 2005, the US FDA promulgated new guidance on the selection and use of excipients in nonclinical and clinical studies. US FDA compliance officials require the use of inactive ingredients that meet compendial standards when standards exist and either have previous use in US FDA approved pharmaceuticals or that they be qualified as "novel" excipients. The FDA/CDER maintains an Inactive Ingredient Committee whose charter includes the evaluation of the safety of inactive ingredients on an as-needed basis, preparation of recommendations concerning the types of data needed for excipients to be declared safe for inclusion in a drug products, and other related functions.

From a regulatory standpoint, FDA's concern regarding safety involves the toxicity, degradants, and impurities of excipients, as discussed in other chapters in this book. In addition, other chapters of this book address types of toxicity concerns, toxicity testing strategies, and exposure and risk assessment of excipients.

The *USP-NF* provides a listing of excipients by categories in a table according to the function of the excipient in a dosage form, such as tablet binder, disintegrant, and such. An excellent reference for excipient information is the *Handbook of Pharmaceutical Excipients* (Rowe *et al.*, 2012). Additionally, Gad *et al.* (2006, 2015) provide an excellent and extensive database of nonclinical formulation components and either acceptable maximum usage levels by species route and duration of study. A new and much expanded version of this formulary has just been submitted for publication.

Excipients have historically not been subjected to extensive safety testing because they have been considered *a priori* to be biologically inactive, therefore, nontoxic. Many, if not most, excipients used are approved food ingredients, the safety of which has been assured by a documented history of safe use or appropriate animal testing. Some of the excipients are generally recognized as safe (GRAS) food ingredients. The excipient is an integral component of the finished drug preparation and, in most countries, is evaluated as part of this preparation. There has been no apparent need to develop specific guidelines for the safety evaluation of excipients, and most developed countries do not have specific guidelines. However, as drug development has become more complex and/or new dosage forms have developed, improved drug bioavailability has become more important. It was noted that the available excipients were often inadequate, new pharmaceutical excipients specifically designed to meet the challenges of delivering new drugs were needed, and these are being developed. The proper safety evaluation of new excipients has now become an integral part of drug safety evaluation.

In the absence of official regulatory guidelines, the safety committees of the IPEC in the United States, Europe, and Japan developed guidelines for the proper safety evaluation of new pharmaceutical excipients (IPEC, 2012). The committees critically evaluated guidelines for the safety evaluation of food ingredients, cosmetics, and other products, as well as textbooks and other appropriate materials. Before initiating a safety evaluation program for a new pharmaceutical excipient, it is advisable to address the following:

1) Chemical and physical properties and functional characterization of the test material
2) Analytical methods that are sensitive and specific for the test material and that can be used to analyze for the test material in animal food used in the feeding studies or in the vehicle used for other studies
3) Available biological, toxicological, and pharmacological information on the test material and related materials (which involves a thorough search of the scientific literature)
4) Intended conditions of use, including reasonable estimates of exposure
5) Potentially sensitive segments of the population

3.4 Techniques of Oral Administration

There are three major techniques for oral delivery of drugs to test animals. The most common way is by gavage, which requires that the material be in a solution or suspension for delivery by tube to the stomach. Less common materials may be given as capsules (particularly to dogs) or in diet (for longer-term studies). Rarely, oral studies may also be done by inclusion of materials in drinking water.

Test materials may be administered as solutions or suspensions as long as they are homogenous and delivery is accurate. For traditional oral administration (gavage), the solution or suspension can be administered with a suitable stomach tube or feeding needle ("Popper" tube) attached to a syringe. If the dose is too large to be administered at one time, it can be divided into equal subparts with 2–4h between each administration; however, this subdivided dosing approach should generally be avoided.

Test chemicals placed into any natural orifice such as the mouth exert local effects and, in many instances, systemic effects as well. The possibility of systemic effects occurring when local effects are to be evaluated should be considered.

For routes of administration in which the chemical is given orally or placed into an orifice other than the mouth, clear instructions about the correct administration of the chemical must be provided. Many cases are known of oral pediatric drops for ear infections being placed into the ear and vice versa (ear drops being swallowed) in humans. Errors in test article administration are especially prevalent when a chemical form is being used in a nontraditional manner (e.g., suppositories that are given by the buccal route).

Administration of a drug in capsules is a common means of dosing larger test animals (particularly dogs). It is labor intensive (each capsule must be individually prepared, though robotic systems are now available for this), but capsules offer the advantages that neat drug may be used (no special formulation need be prepared, and the questions of formulation or solution stability are avoided); the dogs are less likely to vomit, and the actual act of dosing requires less labor than using a gavage tube. Capsules may also be used with primates, though they are not administered as easily. It is also possible to make microcapsules for rodents (Melnick *et al.*, 1987).

Incorporation of a drug in the diet is commonly used for longer-term studies (particularly carcinogenicity studies, though the method is not limited to these). Dosing by diet is much less labor intensive than any other oral dosing methodology, which is particularly attractive over the course of a long (13-week, 1-year, 18-month, or 2-year) study.

The most critical factor to dietary studies is the proper preparation of the test chemical–diet admixtures. The range of physical and chemical characteristics of test materials requires that appropriate mixing techniques be determined on an individual basis. Standard practices generally dictate the preparation of a premix, to which is added appropriate amounts of feed to achieve the proper concentrations.

Dietary preparation involving liquid materials frequently results in either wet feed in which the test article does not disperse or formation of "gumballs"—feed and test material that form discernible lumps and chemical "hot-spots." Drying and grinding of the premix to a free-flowing form prior to mixing the final diets may be required; however, these actions can affect the chemical nature of the test article.

Solid materials require special techniques prior to or during addition to diets. Materials that are soluble in water may be dissolved and added as described earlier for liquids. Non-water-soluble materials may require several preparatory steps. The test chemical may be dissolved in corn oil, acetone, or other appropriate vehicle prior to addition to the weighed diet. When an organic solvent such as acetone is used, the mixing time for the premix should be sufficient for the solvent to evaporate. Some solids may require grinding in a mortar and pestle with feed added during the grinding process.

Prior to study initiation, stability of the test chemical in the diet must be determined over a test period at least equivalent to the time period during which animals are to be exposed to a specific diet mix. Stability of test samples under the conditions of the proposed study is preferable. Labor and expense can be saved when long-term stability data permit mixing of several weeks (or a month) of test diet in a single mixing interval.

Homogeneity and concentration analysis of the test article–diet admixture are performed by sampling at three or four regions within the freshly mixed diet (e.g., samples from the top, middle, and bottom of the mixing bowl or blender).

A variety of feeders are commercially available for rats and mice. These include various-sized glass jars and stainless steel or galvanized feed cups, which can be equipped with restraining lids and food followers to preclude significant losses of feed due to animals digging in the feeders. Slotted metal feeders are designed so that animals cannot climb into the feed, and they also contain mesh food followers to prevent digging.

Another problem sometimes encountered is palatability—the material may taste so strongly that animals will not eat it. As a result, palatability, stability in diet, and homogeneity of mix must all be ensured prior to the initiation of an actual study.

Inclusion in drinking water is rarely used for oral administration of human drugs to test animals, though it sees more frequent use for the study of environmental agents.

Physicochemical properties of the test material should be a major consideration in selecting of drinking water as a dosing matrix. Unlike diet preparation or preparation of gavage dose solutions and suspensions where a variety of solvents and physical processes can be utilized to prepare a dosable form, preparations of drinking water solutions are less flexible. Water solubility of the test chemical is the major governing factor and is dependent on factors such as pH, dissolved salts, and temperature. The animal model itself sets limitations for these factors (acceptability and suitability of pH and salt-adjusted water by the animals as well as animal environmental specifications such as room temperature).

Stability of the test chemical in drinking water under study conditions should be determined prior to study initiation. Consideration should be given to conducting stability tests on test chemical–drinking water admixtures presented to some test animals. Besides difficulties of inherent stability, changes in chemical concentrations may result from other influences. Chemicals with low vapor pressure can volatilize from the water into the air space located above the water of an inverted water bottle; thus, a majority of the chemical may be found in the "dead space," not in the water.

Certain test chemicals may be degraded by contamination with microorganisms. A primary source of these microorganisms is the oral cavity of rodents. Although rats and mice are not as notorious as the guinea pig in spitting back into water bottles, significant bacteria can pass via the sipper tubes and water flow restraints into the water bottles. Sanitation and sterilization procedures for water bottles and sipper tubes must be carefully attended to.

Many technicians may not be familiar with terms such as sublingual (under the tongue), buccal (between the cheek and gingiva), otic, and so on. A clear description of each of these nontraditional routes (i.e., other than gavage routes) should be discussed with technicians, and instructions may also be written down and given to them. Demonstrations are often useful to illustrate selected techniques of administration (e.g., to use an inhaler or nebulizer). Some chemicals must be placed by technicians into body orifices (e.g., medicated intrauterine devices such as Progesterset).

3.4.1 Volume Limitations by Oral Route

In the strictest sense, absolute limitations on how much of a dosage form may be administered by any particular route are determined by specific aspects of the test species or dosage form. But there are some general guidelines (determined by issues of humane treatment of animals, accurate deliver of dose, and such) that can be put forth. These are summarized in Table 3.8. The Appendix and Section 3.3.1 should, of course, be checked to see if there is specific guidance due to the characteristics of a particular vehicle.

3.4.2 Route Comparisons and Contrasts

Though the range of routes of administration of a drug that may be used is beyond the scope of this volume, it is important to realize that route and total extent of systemic absorption are but two aspects of route specific effects on drug effects and toxicity.

The first part of this chapter described, compared, and contrasted the various routes used in toxicology and presented guidelines for their use. There are, however, some exceptions to the general rules that the practicing toxicologist should keep in mind.

Equally important is that vehicles may also have (at least transiently) biologic effects. *Vehicles can mask the effects of active ingredients.* Particularly for clinical signs, attention should be paid to the fact that a number of vehicles (e.g., propylene glycol) cause transient neurobehavioral effects that may mask similar short-lived (though not necessarily equally transient and reversible) effects of test materials.

There are multiple approaches to enhance the bioavailability of candidate molecules, which offer superior therapeutic receptor site specificity and metabolic stability.

The most common approaches are formulation, which is the major theme of this volume. Especially in the early nonclinical phase of development, we want to be able to get our candidates into both *in vitro* (cell culture models where we early on can evaluate potential for mutagenicity, hERG channel inhibition, metabolic stability, protein binding, and such) and *in vivo* (where we can evaluate organism level tolerance, pharmacokinetics, and pharmacodynamics) in model systems where we tend to have less variability between individual model units (cells and animals) and do not have the costs or restraints inherent in both GLP nonclinical and clinical studies.

A second approach is to seek improved bioavailability by developing prodrugs (molecules with little inherent pharmacodynamic activity but highly favorable bioavailability). Such molecules once absorbed are transformed into molecules, which have highly favorable pharmacodynamic profiles.

3.4.3 Calculating Material Requirements

One of the essential basic skills for the efficient design and conduct of preclinical studies is to be able to accurately project compound requirements for the conduct of a study. In theory, this simply requires plugging numbers into a formula such as

$$(A \times B \times C \times D)1.1 = \text{total compound requirement}$$

where:

A = number of animals in each study group.
B = the *sum* of doses of the dose groups (such as $0.1 + 0.3 + 1.0 \text{ mg/kg} = 1.4 \text{ mg/kg}$).
C = the number of doses to be delivered (usually the length of the study in days).
D = the average body weight per animal (assuming dosing is done on a per body weight basis).
1.1 = a safety factor (in effect, 10%) to allow for spillage, and so on.

As an example of this approach, consider a study that calls for 10 dogs/sex/ group ($A = 10 \times 2 = 20$) to receive 0, 10, 50, or 150 mg/kg/day ($B = 10 + 50 + 150 = 210 \text{ mg/kg}$) for 30 days ($C = 30$). On average, our dogs of the age range used to weigh 10 kg ($D = 10 \text{ kg}$). Our compound need is then $(20 \times 210 \text{ mg/} \text{kg} \times 30 \times 10 \text{ kg}) 1.1 = 1.386 \text{ kg}$.

The real-life situation is a bit more complicated, since animal weights change over time, diet studies have doses dependent on daily diet consumption, the material may be a salt but dosage should be calculated on the basis of the parent compound, and not all animals may be carried through the entire study.

For rats and mice (where weight change is most dramatic and diet studies most common), Table 3.11 presents some reliable planning values for compound requirements during diet studies.

3.5 Concluding Remarks

The longer the expected human exposure, the more extensive will be the toxicological studies to assure safety of the drug and of excipients. Toxicity is a result of dose (either C_{max} or an accumulation effect) and duration of exposure.

Table 3.11 Standardized total compound requirements for rodent diet studies.[a]

	Total compound requirement (g) per dose (mg/kg/day)					
Length of study	1	3	10	30	100	300
Rat[b]						
2 weeks	0.2	0.4	1.2	4	10.6	32
4 weeks	0.43	0.7	2.5	7.5	25	75
13 weeks	0.8	2.6	8.5	25.5	85	260
52 weeks	7	21	70	210	0.7[c]	2.1[c]
2 years	15	45	150	450	1.5[c]	4.5[c]
Mouse						
2 weeks	0.03	0.06	0.22	0.65	2.2	6.4
4 weeks	0.08	0.14	0.8	1.4	8	14
13 weeks	0.14	0.42	1.4	4.2	14	42
18 months	0.85	2.5	8.5	25	85	250

[a] Based on 10 animals per sex per group for the length of the study that are 6–8 weeks old at study initiation. Animals are weighed to determine body weights.
[b] Sprague Dawley rats (body weights and compound requirements for Fischers would be less).
[c] In kilograms.

A tiered approach assures that those tests necessary to ensure safety for the expected duration of human exposure are conducted. Thus, to assure safe use for greater than 2 weeks, but no more than 6 weeks in humans, subchronic toxicity and developmental toxicity studies are required. This means long-term studies should be considered for prolonged human exposures, but may not be absolutely required. From a critical evaluation by a competent toxicologist, the results of the physical–chemical properties of the test material, the 28- and 90-day tests, the ADME–PK acute and repeated dose tests, and the developmental toxicity test(s), a final determination, can be made on the value of chronic toxicity or oncogenicity studies.

For example, if no toxicity is observed at a limit dose of 1000 mg/kg body weight/day following the 90-day toxicity study, no genotoxicity was found, and the ADME–PK profile indicates that the material is not absorbed and is completed excreted unchanged in the feces, then it is likely that a chronic study is not necessary. The decision to conduct chronic studies should be determined on a case-by-case basis using scientific judgment with the scheme and requirements evolving over the course of the ICH globalization process.

The solubility requirements for potential oral drugs during preclinical safety studies conducted before the main line GLP general toxicity studies operate along different lines than those "just before man" (IND-enabling) studies.

For two of the three groups of studies here (genotoxicity and safety pharmacology), the primary objective is simply to achieve sufficient solubility in the test model environment (cell or bacterial culture) to fulfill the requirements of ICH exposure guidelines. This means that solvents such as DMSO, NMP, or hexane may be utilized, as no other systemic toxicity or regulatory limits apply.

For the studies that do involve intact animals at this stage (mouse micronucleus, cardiovascular, or respiratory safety), only single dose is generally to be administered. The principal limitation of the single dose is avoiding acute lethality. So again, a wide range of vehicles is available.

Only when we progress to the pilot systemic toxicity studies (MTD and DRF now commonly performed in a single two-phase study using only a few animals) do we become more limited by the need to have at least short-term (up to 7 days) tolerance, avoiding target organ toxicities (Brown, 1980; Brown and Muir, 1971; Gad *et al.*, 1984).

Properly done, the results of this process provide a great deal of not only pharmaco- and toxicodynamic data but also significant information to guide the development of more sophisticated formulations with which to progress to IND and then to humans. In so doing, as illustrated in Figure 3.5, we seek to maximize the range of the dose–response curve, which is therapeutically quite effective in (the anticipated effect) minimizing the incidence of toxic effects.

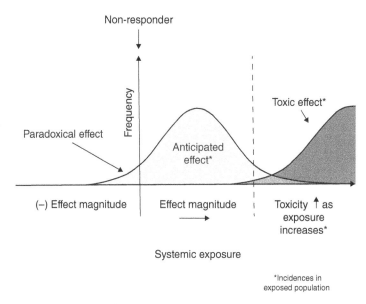

Figure 3.5 Dose–response curve spectrum (objective is to maximize the range of the anticipated (desired) effect zone).

References

Adeyeye, M. C., and Brittain, H. G. (Eds.) (2008). *Preformulation in Solid Dosage Form Development*. In: Drugs and the Pharmaceutical Sciences, Vol. **178**, Informa Healthcare USA, New York.

APA (1994). *Handbook of Pharmaceutical Excipients*, APA, Chicago.

Baker, H. J., Lindsey, J. R., and Weisbroth, S. H. (1979). *The Laboratory Rat*, Vol. **1**. Academic Press, New York, pp. 411–412.

Bates, T. R., and Gibaldi, M. (1970). Gastrointestinal absorption of drugs. In: *Current Concepts in the Pharmaceutical Sciences: Biopharmaceutics* (Swarbrick, J., Ed.), Lea & Febiger, Philadelphia, pp. 58–99.

Batrakova, E. V., Li, S., and Miller, D. W. (1999). Pluronic P85 increases permeability of a broad spectrum of drug in polarized BBMEC and CaCo2 cell monolayer. *Pharm. Res.* **16**:1366.

Boersen, N., Lee, T., and Hui, H.-W. (2014). Development of preclinical formulations for toxicology studies. In: *A Comprehensive Guide to Toxicology in Preclinical Drug Development* (Faqi, A. S., Ed.), Academic Press, San Diego, pp. 69–86.

Borowitz, J. L., Moore, P. F., Yim, G. K. W., and Miya, T. S. (1971). Mechanism of enhanced drug effects produced by dilution of the oral dose. *Toxicol. Appl. Pharmacol.* **19**:164–168.

Brown, V. K. (1980). *Acute Toxicity in Theory and Practice*, John Wiley & Sons, Inc., New York.

Brown, V. K., and Muir, C. M. C. (1971). Some factors affecting the acute toxicity of pesticides to mammals when absorbed through skin and eyes. *Int. Pest Control* **13**:16–21.

Caldwell, G. W., Ritchie, D. M., Masucci, J. A., Hageman, W., and Yan, Z. (2001). The new pre-preclinical paradigm: compound optimization in early and late phase drug discovery. *Curr. Top. Med. Chem.* **1**(5):353–366.

Calvary, H. O., and Klump, T. G. (1939). The toxicity for human beings for diethylene glycol with sulfanilamide. *South. Med. J.* **32**:1105.

Fitzgerald, J. M., Boyd, V. F., and Manus, A. G. (1983). Formulations of insoluble and immiscible test agents in liquid vehicles for toxicity testing. In: *Chemistry for Toxicity Testing* (Jameson, C. W., and Walters, D. B., Eds.), Butterworth, Boston, pp. 83–90.

Gad, S. C. (Ed.) (2005). *Drug Discovery Handbook*, John Wiley & Sons, Inc., Hoboken.

Gad, S. C. (Ed.) (2008a). *Preclinical Development Handbook: ADME and Biopharmaceutical Properties*, John Wiley & Sons, Inc., Hoboken.

Gad, S. C. (2008b). *Pharmaceutical Manufacturing Handbook: Production and Processes*, John Wiley & Sons, Inc., Hoboken.

Gad, S. C. (Ed.) (2015). *Animal Models in Toxicology*, 3rd Edition, CRC Press, Boca Raton.

Gad, S. C. (2016). *Drug Safety Evaluation*, 3rd Edition, John Wiley & Sons, Inc., Hoboken.

Gad, S. C., Smith, A. C., Cramp, A. L., Gavigan, F. A., and Derelanko, M. J. (1984). Innovative designs and practices for acute systemic toxicity studies. *Drug Chem. Toxicol.* 7(5):423–434.

Gad, S. C., Cassidy, C., Aubert, N., Spainhour, B., and Robbe, H. (2006). Nonclinical vehicle use in studies by multiple routes in multiple species. *Int. J. Toxicol.* 25:499–522.

Gad, S. C., Spainhour, B., Shoemaker, C., Pallman, D. R. A., Stricker-Krongrad, A., Downing, A., Seals, R. E., Eagle, L., Polhamus, K., and Daly, J. (2015). Tolerable levels of nonclinical vehicles and formulations used in studies by multiple routes in multiple species with notes on methods to improve utility, *Int. J. Toxicol.* 35(2):95–178.

Gaines, T. B. (1960). The acute toxicity of pesticides to rats. *Toxicol. Appl. Pharmacol.* 2:88–99.

Gibaldi, M. (1976). Biopharmaceutics. In: *Theory and Practice of Industrial Pharmacy*, 2nd Edition (Lachman, L., Lieberman, H. A., and Kanig, J. L., Eds.), Lea & Febiger, Philadelphia.

Gibaldi, M., and Feldman, S. (1970). Mechanisms of surfactant effects on drug absorption. *J. Pharm. Sci.* 59:579.

Gibson, M. (Ed.) (2009). *Drugs and the Pharmaceutical Sciences, Pharmaceutical Preformulation and Formulation*, 2nd Edition, Informa Healthcare USA, New York, pp. 172–187.

Gough, W. B., Zeiler, R. H., Barreca, P., and El Sherif, N. (1982). Hypotensive action of commercial intravenous amiodarone and polysorbate 80 in dogs. *J. Cardiovasc. Pharmacol.* 4(3):375–380.

Groves, M. (1966). The influence of formulation upon the activity of thermotherapeutic agents. *Reps. Progr. Appl. Chem.* 12:51–151.

Hageman, M. J. (2010). Preformulation designed to enable discovery and assess developability. *Comb. Chem. High Throughput Screen.* 13:90–100.

Hagenbusch, O. E. (1937). Elixir of Sulfanilamide–Massengill. *J. Am. Med. Assoc.* 109:1531.

Hardee, G. E., and Baggo, J. D. (1998). *Development and Formulation of Veterinary Dosage Forms*, 2nd Edition. Drugs and the Pharmaceutical Sciences, Vol. 88, CRC Press, New York.

Higgins, J., Cartwright, M. E., and Templeton, A. C. (2012). Foundation review: progressing preclinical drug candidates: strategies on preclinical safety studies and the quest for adequate exposure. *Drug Discov. Today* 17(15–16):828–836.

Hogben, C. A. M., Tocco, D. J., Brodie, B. B., and Schranker, L. S. (1959). On the mechanism of intestinal absorption of drugs. *J. Pharmacol. Exp. Ther.* T25:275.

Holick, H. G. O. (1937). Glycerine, ethylene glycol, propylene glycol and diethylene glycol. *J. Am. Med. Assoc.* 109:1517.

Houston, J. B., Upshall, D. G., and Bridges, J. W. (1974). The re-evaluation of the importance of partition coefficients in the gastrointestinal absorption of nutrients. *J. Pharmacol. Exp. Ther.* **189**:244–254.

IPEC (2012). The Joint IPEC–PQG Guide for Pharmaceutical Excipients, Pharmaceutical Quality Group, London.

Katdare, A., and Chaubal, M. V. (2006). *Excipient Development for Pharmaceutical Biotechnology and Drug Delivery Systems*, Informa Healthcare, New York.

Kerberle, H. (1971). Physicochemical factors of drugs affecting absorption, distribution and excretion. *Acta Pharmacol. Toxicol.* **29**(Suppl 3):30–47.

Kwong, E. (2015). Advancing drug discovery: a pharmaceutics perspective. *J. Pharm. Sci.* **104**:865–871.

Kwong, E., Higgins, J., and Templeton, A. C. (2011). Strategies for bringing drug delivery tools into discovery. *Int. J. Pharm.* **412**:1–7.

Le-Ngoc Vo, C., Park, C., and Beom-Jin, L. (2013). Current trends and future perspectives of solid dispersions containing poorly water-soluble drugs. *Eur. J. Pharm. Biopharm.* **85**:799–813.

Lien, E., Koda, R. T., and Tong, G. L. (1971). Buccal and percutaneous absorptions. *Drug Intell. Clin. Pharm.* **5**:38–41.

Lipinski, C. A. (2004). Lead- and drug-like compounds: the rule-of-five revolution. *Drug Discov. Today Technol.* **1**(4):337–341.

Lipinski, C. A., Lombardo, F., Dominy, B. W., and Feeney, P. J. (2001). Experimental and computational approaches to estimate solubility and permeability in drug discovery and development settings. *Adv. Drug Deliv. Rev.* **46**(1–3):3–26.

Liu, R. (Ed.) (2008). *Water-Insoluble Drug Formulation*, 2nd Edition. CRC Press, Boca Raton.

Lohani, S., Cooper, H., Jin, X., Nissley, B., Manser, K., Rakes, L. H., Cummings, J. J., Fauty, S. E., and Bak, A. (2014). Physiological properties, form, and formulation selection strategy for biopharmaceutical classification system class II preclinical drug candidate. *J. Pharm. Sci.* **103**:3007–3021.

Maas, J., Kamm, W., and Hauck, G. (2007). An integrated early formulation strategy—from hit evaluation to preclinical candidate profiling. *Eur. J. Pharm. Biopharm.* **66**:1–10.

Melnick, R. L., Jameson, C. W., Goehl, T. J., and Kuhn, G. O. (1987). Application of microencapsulation for toxicology studies. *Fundam. Appl. Toxicol.* **11**:42531.

Monkhouse, D. C., and Rhodes, C. T. (1998). *Drug Products for Clinical Trials*, Marcel Dekker, New York.

Mountfield, R., Senepin, S., Schleimer, M., Walter, I., and Bittner, B. (2000) The potential inhibitory effect of formulation ingredients on intestinal cytochrome P450. *Int. J. Pharm.* **211**:89–92.

Neervannan, S. (2006). Preclinical formulations for discovery and toxicology: physicochemical challenges. *Expert Opin. Drug Metab. Toxicol.* **2**(5):715–731.

Newman, A., Knipp, G., and Zografi, G. (2012). Assessing the performance of amorphous solid dispersions. *J. Pharm. Sci.* **101**(4):1355–1377.

Niazi, S. K. (2007). *Handbook of Preformulation: Chemical, Biological, and Botanical Drugs*, Informa Healthcare USA, New York.

Palucki, M., Higgins, J. D., Kwong, E., and Templeton, A. C. (2010). Strategies at the interface of drug discovery and development: early optimization of the solid state phase and preclinical toxicology formulation for potential drug candidates. *J. Med. Chem.* **53**(16):5897–5905.

Rabinow, B. E. (2004). Nanosuspensions in drug delivery. *Nat. Rev. Drug Discov.* **3**:785–796.

Racy, I. (1989). *Drug Formulation*, John Wiley & Sons, Inc., New York.

Ritschel, W. A., Siegel, E. G., and Ring, P. E. (1974). Biopharmaceutical factors influencing LDP: Pt. 1. Viscosity. *Arztl. Forsch.* **24**:907–910.

Rowe, R. C., Sheskey, P. J., Cook, W. G., and Fenton, M. E. (2012). *Handbook of Pharmaceutical Excipients*, 7th Edition, American Pharmaceutical Association, Washington, DC.

Schranker, L. S. (1960). On the mechanism of absorption of drugs from the gastrointestinal tract. *J. Med. Pharm. Chem.* **2**:343–359.

Schulze, J. D. R., Waddington, W. A., Ell, P. J., Parsons, G. E., Coffin, M. D., and Basit, A. W. (2003). Concentration-dependent effects of polyethylene glycol 400 on gastrointestinal transit and drug absorption. *Pharm. Res.* **20**(12):1984–1988.

Selassie, C., and Verma, R. P. (2010). History of quantitative structure–activity relationship. In: *Burgers Medicinal Chemistry, Drug Discovery and Development* (Abraham, D. J., and Rotella, D. P., Eds.), John Wiley & Sons, Inc., Hoboken.

Shah, A. K., and Agnihotri, S. A. (2011). Recent advances and novel strategies in pre-clinical formulation development: an overview. *J. Control. Release* **156**:281–296.

Smolinske, S. C. (1992). *Handbook of Food, Drug and Cosmetic Excipients*, CRC Press, Boca Raton.

Sotthivirat, S., McKelvey, C., Moser, J., Rege, B., Xu, W., and Zhang, D. (2013). Development of amorphous solid dispersion formulations of a poorly water-soluble drug, MK-0364. *Int. J. Pharm.* **452**:73–81.

Stahl, P. H., and Wermuth, C. G. (Eds.) (2011). *Handbook of Pharmaceutical Salts: Properties, Selection and Use*, 2nd Edition, Wiley-VCH & VHCA, Zurich.

Strickley, R. G. (2008). Formulation in drug discovery. In: *Annual Report on Medicinal Chemistry* (Wood, A., and Desai, M., Eds.), Elsevier, Oxford, pp. 419–451.

Strickley, R. G. (2011). Formulations in Drug Discovery: Enabling Preclinical Pharmacologic Pharmacodynamic and Toxicology Studies. Bay Area Discussion Group, Gilead Sciences, Inc., December 1, 2011.

Swarbrick, J. (Ed) (2006) *Current Concepts in Pharmaceutical Sciences: Dosage Form, Design and Bioavailability*, Lea & Febiger, Philadelphia.

Thackaberry, E. A. (2012). Non-clinical toxicological considerations for pharmaceutical salt selection. *Expert Opin. Drug Metab. Toxicol.* **8**(11): 1419–1433.

Thackaberry, E. A. (2013). Vehicle selection for nonclinical oral safety studies. *Expert Opin. Drug Metab. Toxicol.* **9**(12):1635–1646.

USFDA (2005) Guidance for Industry: Nonclinical Studies for the Safety Evaluation of Pharmaceutical Excipients. Available at http://www.fda.gov/cder/guidance/5544fn1.pdf (accessed February 2009).

USP-NF (2014). United States Pharmacopeia 37–National Formulary 32, United States Pharmacopiea Convention, Inc., Washington, DC.

Van Zuylen, I., Verweij, J., and Sparebrook, A. (2001). Role of formulation vehicles in taxane pharmacology. *Invest. New Drugs* **19**:125–141.

Wagner, D., Spahn-Langguth, H., Hanafy, A., Koggel, A., and Langguth, P. (2001). Intestinal drug efflux: formulation and food effects. *Adv. Drug Deliv. Rev.* **50**(Suppl 1):S13–S31.

Wiener, M. L., and Katkoskie, L. A. (1999). *Excipient Toxicity and Safety*, Marcel Dekker, Inc., New York.

Wilson, F. A., and Dietschy, J. M. (1974). The intestinal unstirred layer—its surface area and effect on active transport kinetics. *Biochim. Biophys. Acta* **363**:112–126.

Winne, D. (1978). Dependence of intestinal absorption *in vivo* in the unstirred layer. *Naunyn Schmiedeberg's Arch. Pharmacol.* **304**:175–181.

Wood, S. G., Upshall, D. G., and Bridges, J. W. (1978). The absorption of aliphatic carbamates from the rat colon. *J. Pharm. Pharmacol.* **30**:638–641.

Zheng, W., Jain, A., Papoutsakis, D., Dannenfelser, R.-M., Panicucci, R., and Garad, S. (2012). Selection of oral bioavailability enhancing formulations during drug discovery. *Drug Dev. Ind. Pharm.* **38**(2):235–247.

4

Bridging End of Discovery to Regulatory Filing: Formulations for IND- and Registration-Enabling Nonclinical Studies

Evan A. Thackaberry

Genentech, Inc., Safety Assessment, South San Francisco, CA, USA

4.1   Introduction

Once lead optimization and pilot toxicology studies are completed, more intense development activities take over in preparation of good laboratory practice (GLP) toxicology studies and GMP manufacturing of API and drug product. During this stage of development, additional data is generated in order to characterize the API phase, formulation requirements, pharmacokinetics,

Oral Formulation Roadmap from Early Drug Discovery to Development,
First Edition. Edited by Elizabeth Kwong.
© 2017 John Wiley & Sons, Inc. Published 2017 by John Wiley & Sons, Inc.

metabolism, distribution, and safety of the lead molecule. The ultimate goal of this effort is to support phase 1 clinical trials in humans. This chapter will examine the interplay of formulation strategy and the GLP studies to support the candidate progressing into clinical development.

4.2 Formulation Selection for GLP Nonclinical Safety Studies

As the development of a small molecule new chemical entity (NCE) approaches Investigational New Drug (IND) application-enabling safety studies, the choice of the nonclinical formulation becomes more critical, and the options available become more limited. There are numerous reasons for this, including the requirement for high doses in safety studies, the large dose volumes required to achieve these doses, and the sub-chronic/chronic nature of the dosing. The minimum requirements for a formulation supporting IND-enabling safety studies are given in Table 4.1. Each of these requirements will be discussed in detail later in this chapter, and each presents a significant challenge. As a result of these requirements, many formulations that are useful for efficacy and single-dose pharmacokinetics/tolerability studies are no longer acceptable for IND-enabling safety studies. However, the most limiting factor is often the maximal achievable dose concentration and associated exposure, which is critical for adequate safety assessment of an NCE.

In order to fully understand formulation requirements for GLP toxicology studies, it is critical to understand the goals of these studies and the regulations that guide study selection and design. The nonclinical testing paradigm for NCEs is dictated to a large degree by the indication and intended duration of clinical exposure. For NCEs intended to be dosed for less than 6 consecutive months, the duration of the nonclinical toxicology studies is generally similar to the intended clinical dosing duration. For NCEs intended to be given chronically (>6 months), long-term chronic toxicology studies (typically 6–9 months) and carcinogenicity studies are generally warranted (ICH, 1995, 2009a). NCEs intended as first-line therapies for near-term life-threatening oncology

Table 4.1 Formulation requirements for nonclinical safety studies.

---

1) Produces adequate exposure to the NCE
2) Allows for technically feasible repeat-dose administration
3) Allows for sufficient stability and homogeneity of the test article
4) Tolerated by nonclinical species under the condition of the study
5) Does not interfere with the ability to assess the safety of the test article under the conditions of the study

---

indications and occasionally other rapidly terminal illnesses are held to a different standard in terms of the safety studies and maximal doses required (ICH, 2009b). Typically, repeat-dose studies of only 3 months are sufficient for registration of these drugs, and carcinogenicity studies are not required. It is important to note, however, that not all oncology indications will be treated in such a way by health authorities. Adjuvant therapies (non-first line) or NCEs intended to treat well-controlled neoplasias with numerous therapeutic alternatives may still be held to a non-oncology standard. As the number of treatment options for oncology indications increases, the number of oncology drugs being held to non-oncology nonclinical testing standards also increases.

There is no "standard" set of nonclinical studies to support all clinical trials. Instead, the guidances put forward by the Food and Drug Administration (FDA) and the International Committee on Harmonization (ICH) offer a framework for determining which studies will be required to support a given clinical trial or intended use based on the indication, known pharmacology and toxicology, clinical dose levels, route, and frequency and duration of administration. For the purposes of this chapter, we are referring mainly to oral small molecule drugs, which eliminates some studies, such as tissue cross-reactivity (for therapeutic antibodies) and local tolerance (for non-oral drugs). However, there will still be quite a bit of variability in the studies run from program to program. Table 4.2 provides a high level description of typical oral nonclinical toxicology programs for oncology and non-oncology indications.

4.2.1 General Toxicology Studies

General toxicology studies are designed to assess the toxicity of an NCE to normal adult animals for a specified duration of exposure. These studies typically include in-life clinical observation, body weight and food consumption assessments, exposure assessment (toxicokinetics), and clinical and anatomic pathology. Specialty biomarkers (serum, urinary, histological, or other) may be added based on known or expected toxicities or for assessment of pharmacodynamic effects. In addition, genotoxicity and safety pharmacology endpoints may be added to these studies in order to compliment standard genotoxicity or safety pharmacology assessments or in some cases to satisfy these regulatory requirements entirely (see Sections 4.2.2 and 4.2.3). Typically, three dose levels are tested with approximately 10 animals/sex/group for rodents and 3–5 animals/sex/group for non-rodents with additional animals added for recovery groups in order to assess the reversibility of the findings. If the intended patient population is entirely a single sex (e.g., oral contraceptives, benign prostate hyperplasia therapies), a single sex may be used for general toxicology studies.

ICH guidances require general toxicology testing in two species, one rodent and one non-rodent, for most oral small molecules. While large molecules such as therapeutic antibodies often assess safety in only one species due to the

Table 4.2 Typical safety studies to support clinical trials and registration of an oral small molecule NCE.

Stage of development	Non-oncology/adjuvant therapy	Oncology[a]
IND-enabling studies	*In vitro* bacterial reverse gene mutation assay (Ames)	*In vitro* bacterial reverse gene mutation assay (Ames)
	Mammalian cell genotoxicity (*in vitro* and/or *in vivo*)	Mammalian cell genotoxicity (*in vitro* and/or *in vivo*)
	In vitro hERG assay	*In vitro* hERG assay
	Rodent repeat-dose GLP general toxicology	Rodent repeat-dose GLP general toxicology
	Non-rodent repeat-dose GLP general toxicology	Non-rodent repeat-dose GLP general toxicology
	CV and CNS safety pharmacology[b]	CV and CNS safety pharmacology[b]
"Large" clinical trials[c]	Phototoxicity (*in vitro* and/or *in vivo*)	Phototoxicity (*in vitro* and/or *in vivo*)
	Rodent teratogenicity study	Rodent teratogenicity study[d]
	Rabbit teratogenicity study	Rabbit teratogenicity study[d]
Chronic dosing (>3 months)	Fertility and early embryonic developmental toxicity study	Fertility and early embryonic developmental toxicity study[d]
	Rodent 6-month repeat-dose general toxicology	
	Non-rodent 9-month repeat-dose general toxicology	
Registration	Perinatal–postnatal toxicity	
	Mouse carcinogenicity	
	Rat carcinogenicity	

[a] NCEs for immediately life-threatening indications only.
[b] Some or all safety pharmacology assessments may be rolled into repeat-dose toxicology studies.
[c] Includes studies with chronic (>6 months) clinical administration or shorter-term studies including more than 150 women of childbearing potential.
[d] For immediately life-threatening indications with NCEs expected to be in development or reproductive toxicants, pilot teratogenicity studies in one species with clear teratogenicity observed is sufficient.

requirement for pharmacologic activity, even in the rare instance when an oral small molecule is pharmacologically active in only one nonclinical species, two species are still assessed in order to address the risk of off-target toxicities. The selection of species is generally dependent on metabolic profile as compared to the predicted human metabolism, pharmacologic activity, and species sensitivity, with the most sensitive species generally used, unless the sensitivity of this species is known to not be predictive of the human toxicology profile. Rat is used for the vast majority of oral small molecule programs, with dog as the preferred non-rodent species.

As mentioned earlier, duration for each toxicology study must meet or exceed the intended clinical use. For NCEs intended for short-term use, this allows for a relatively simple development plan. For instance, an NCE intended to be administered for only 1 month in the clinic requires only 1 month toxicology studies in two species. For NCEs intended to be dosed over a longer period of time, particularly for those over greater than 6 months, several toxicology studies of increasing duration are generally performed. For example, 1-month toxicology studies may support the phase 1a single ascending dose (SAD) as well as the phase 1b multiple ascending dose (MAD) of up to 1 month of clinical dosing. Three-month toxicology studies may then be performed to support a phase 2 proof-of-concept study with up to 3 months of dosing, followed by 6/9-month chronic studies to support more chronic dosing in phase 3. Similar phase-appropriate toxicology programs are common, since they allow for more rapid entry into the clinic (with shorter IND-enabling studies) while more efficiently utilizing resources and generating data in earlier studies that allow for more informed dose selection for the chronic studies.

The purpose of nonclinical safety studies is to define the toxicity of an NCE across a spectrum of doses in order to assess human risk in the clinic. A key outcome of these studies is the definition of a no observed adverse effect level (NOAEL), which is defined for each study as the dose that causes no toxicologically significant (adverse) effects. The relationship between this dose and the predicted or observed exposures in the clinic is then described as an exposure multiple (also called a safety factor), which is simply the ratio of the exposure (generally area under the curve (AUC)) at the NOAEL in the nonclinical toxicology study and the clinical exposure. For instance, if the NOAEL in a nonclinical toxicology study is associated with an exposure of $100\,\mu M\,h$ and the anticipated efficacious dose in the clinic is $10\,\mu M\,h$, then the exposure multiple is 10×. There is no standardized requirement for exposure multiples in order to proceed to the clinic, though for non-immediately life-threatening indications, an exposure multiple of at least 10× in both nonclinical toxicology species is preferred. The type of toxicity observed above the NOAEL has a significant impact on the risk assessment and acceptability of the exposure multiples as well. For instance, while there may be little concern if the toxicities observed at higher doses were transient, minimal, and reversible, with decreases in hematology parameters, there may be significant concern if the toxicity seen at higher doses was significant and irreversible, such as mortality or target organ necrosis. Because NCEs intended to treat immediately life-threatening neoplasias are often significantly more toxic than non-oncology compounds, the NOAELs and maximal doses in these studies tend to be much lower, generally only 2–10× the predicted therapeutic doses, and at times only achieve sub-therapeutic exposure in some more sensitive species. A NOAEL may not be identified in such studies, at which point the nature of the toxicities observed, along with their reversibility and clinical monitorability that become critical for human risk assessment.

Besides the NCE, other components of a formulation may also be qualified in nonclinical safety studies. For non-oral therapeutics, the clinical formulation should be tested nonclinically, and this data can serve to support clinical trials and registration of the excipients therein. Additional discussion about the regulatory path for novel excipients is given in Section 4.7.1. More commonly, nonclinical safety studies are used to qualify NCE impurities (ICH, 2006). Typically, the early GLP batches tested in the nonclinical studies should contain higher levels of such impurities as compared to later GMP batches. This, together with the high dose levels used in these early safety studies, makes them ideal for establishing a safety margin for impurities. This common practice can eliminate the need for additional nonclinical studies later in development.

4.2.2 Genotoxicity Studies

Many of the workhorse genotoxicity assays are performed *in vitro*, which eliminates the need for significant nonclinical formulation work. These include the bacterial reverse mutagenicity assay (the Ames assay), which assesses the potential of an NCE to cause direct damage to DNA (mutagenicity), the chromosomal aberrations, *in vitro* micronucleus, and the mouse lymphoma gene mutation assays, which assess the potential of an NCE to cause chromosomal damage (clastogenicity). However, *in vivo* genotoxicity studies are required before large-scale clinical trials with novel small molecule therapeutics and may be performed prior to IND submission in many cases. The most common *in vivo* assays include the rodent micronucleus and comet assays, which are typically run at high dose levels (maximum tolerated dose (MTD) up to 2 g/kg) using the clinical route of administration and may run as single-dose studies or short repeat-dose studies, depending on the pharmacokinetics of the compound (ICH, 2011). In some cases, they may be rolled into the general toxicology studies in order to reduce animal use.

4.2.3 Safety Pharmacology Studies

Safety pharmacology studies are designed to assess the potential of an NCE to interfere with the normal pharmacology of key organ systems. Such studies may specifically address the effects on the cardiovascular, gastrointestinal (GI), renal, respiratory, or central nervous systems (CNS). In practice, only the "core battery" of cardiovascular system, CNS, and respiratory system is routinely required to support clinical studies, with GI and renal safety pharmacology studies run only for cause. Furthermore, these assessments may be rolled into the general toxicity studies in order to reduce animal use. This is particularly common for respiratory safety pharmacology assessments, which may be accomplished by monitoring oxygen saturation, though cardiovascular and CNS endpoints may also be assessed on general toxicology studies as well. When run as stand-alone studies, these assessments are run as single-dose studies using the clinical route of administration with the high dose producing moderate toxicity (ICH, 2000a). Rodents are commonly used for CNS and respiratory system studies, while non-rodents are used for cardiovascular assessments.

4.2.4 Developmental and Reproductive Toxicology (DART) Studies

The assessment of developmental and reproductive toxicity is required for all small molecule NCEs and may involve a number of specialized study designs and multiple test species. In general, these studies can be classified into three categories: assessment of fertility and early embryonic development, assessment of embryo–fetal development (teratogenicity), and assessment of pre- and postnatal development (ICH, 2000b). These assessments are generally performed as separate studies using distinct endpoints because it is now understood within the field that these distinct windows of reproduction and development are biologically and physiologically distinct and are affected in different ways by a diverse set of toxicants. For instance, compounds that affect the production of germ cells or maintenance of normal hormone cyclicity can impair fertility, while cytotoxic compounds are more likely to be toxic to the early embryo. Traditional teratogens such as thalidomide and valproic acid affect the embryo during the period of organogenesis, while more general feto-toxic compounds or drugs that impair lactation affect the perinatal–postnatal period of development.

Typically, an initial assessment of effects on fertility is achieved by histopatho-logical assessment of the male and female reproductive organs in the general toxicology studies. Therefore, dedicated fertility and early embryonic toxicity studies, which are generally run in a single rodent species, are not generally required until the start of phase 3 clinical trials. Similarly, perinatal–postnatal toxicity studies in a single rodent species are generally only required for market authorization. However, because of the higher risk and more significant toxicities observed, an assessment of embryo–fetal toxicity (teratogenicity) is required in two species (generally rodents and rabbits) prior to any large-scale clinical trials including more than 150 women of childbearing potential. In all cases, these studies are run via the intended clinical route of administration with the top dose expected to cause mild to moderate maternal toxicity. For molecules expected to be teratogenic, such as cytotoxic compounds intended to treat immediately life-threatening cancer indications, demonstration of teratogenicity in a single species pilot toxicology study is generally sufficient, and no further GLP DART studies are required.

4.2.5 Carcinogenicity Studies

An assessment of carcinogenic potential is required for all small molecule NCEs intended for chronic administration of greater than 6 months in duration (ICH, 1995). Additionally, drugs intended to be dosed intermittently but frequently over the course of years will also require carcinogenicity assessment. These studies are long in duration, essentially equivalent to a lifetime exposure in rats and mice (2 years). In recent years, transgenic mouse models have become acceptable, which shortens the study duration to 6 months; however, the 2-year rat bioassay is still required. These studies are also quite large, with 50–75 animals/sex/group, and therefore require a significant quantity of test article.

4.2.6 Other Common GLP Toxicology Studies

Nonclinical development programs are tailored to address the risks associated with a particular NCE. In many cases additional *in vivo* studies may be required beyond those mentioned previously. For example, *in vivo* phototoxicity studies may be performed in rodents for molecules with distinct physiochemical characteristics and *in vitro* phototoxic signals (ICH, 2013). Juvenile toxicity studies may be performed, generally in rodents in order to support clinical use in pediatric populations. Finally, abuse liability studies may be required for potentially CNS-active molecules in order to understand the likelihood of abuse of these drugs. Abuse liability studies will be discussed in more detail later due to their unique formulation requirements.

4.3 Dose Selection and Test Article Requirements for Nonclinical Toxicology Studies

It is the job of the nonclinical toxicologist to define the toxicological profile of an NCE in order to provide a useful human health risk assessment for the clinical trials. An informative risk assessment is difficult to produce unless toxicity is identified in the nonclinical safety studies, since the toxicities observed define the nature of the risk. As such, high doses are often necessary in these safety studies. The major world health agencies (the United States, EU, Japan) have codified a set of guidelines for selection of an appropriate high dose in safety studies with NCEs. These guidances define an acceptable high dose as the MTD, the maximum feasible dose (MFD), the dose producing the maximal exposure, the dose producing a 50× exposure multiple over the anticipated efficacious clinical dose, or (in some cases) a 1000 mg/kg dose (ICH, 2009a).

The MTD is relatively easy to define based on the results of acute and subchronic safety studies. Mortality or significant morbidity that would preclude longer-term dosing (such as body weight loss or serious target organ toxicity) is sufficient to define the MTD. It is important to note that the MTD is typically defined on a per study basis. That is, the MTD is defined under the conditions of a given study. The duration, species, and details of the study design can significantly impact the MTD. For example, for a therapeutic area (TA) with an MTD of 100 mg/kg in a 1-month rat study, the MTD in a 3-month rat study would be expected to be somewhat lower due to the cumulative toxicity of the TA. If the same study is dosed weekly instead of daily, the MTD would be expected to rise significantly. The MTD in other species (such as dogs) is empirically determined in a separate study and is often not predicted by rodent data.

The maximum formulatable dose (MFD) is simply the maximum dose that can be technically achieved due to solubility and/or dose volume limitations. The MFD is more difficult to define for a given NCE, since a sponsor must demonstrate to the health authorities that they are unable to achieve higher doses using any acceptable formulation. For orally administered NCEs, the MFD is often defined by the maximal dosable homogenous concentration obtainable in suspension formulations dosed at the highest tolerated dose volume in a given species. For enabling formulations, the MFD may be defined by the maximal use of enabling vehicles without unwanted toxicological or biologic effects. The assessment of MFD in alternative formulations may be required by health authorities, particularly if no significant toxicity has been identified at the achievable doses with more standard formulations.

Often, absorption-limited compounds reach a plateau in plasma exposure above which higher oral doses fail to produce additional exposure. When this occurs, higher dose levels provide only higher local NCE concentration in the GI tract, which is not useful for human risk assessment. In these cases, the assessment of multiple formulations in an attempt to overcome the plateau in exposure may be required in order to gain endorsement from health authorities on a maximal dose in chronic safety studies. For investigative work in support of defining an MFD or plateau in exposure, TAs do not typically require a large-scale stochastic formulation screening program. Instead, a data-driven approach investigating a few formulations that may be expected to impact exposure, such as corn oil for lipophilic molecules, is generally sufficient.

For most NCEs, the maximal dose that will need to be assessed in nonclinical safety studies is 1000 mg/kg. The only exception to this is for NCEs with intended clinical doses of greater than 1 g/day (not body weight adjusted) in the clinic. In this case, the maximal dose for nonclinical studies is 2000 mg/kg. Due to the large pill burden required for such molecules, doses in this range are typically only acceptable for anti-infective drugs.

The FDA's GLP guidelines are a set of rules (CFR, 1978) designed to ensure the quality of the data produced in nonclinical safety studies used to support the clinical testing of products regulated by the FDA, including drugs and medical devices. Similar guidelines exist in other countries, including the

European Organization for Economic Cooperation and Development (OECD) guidelines. The FDA and other health authorities hold all pivotal (IND- or NDA/BLA-enabling) safety studies to this standard, conduct regular and directed inspections of the testing facilities that conduct these studies, and are able to disqualify studies or testing facilities if the GLP guidelines are not followed. The GLPs are designed to ensure the quality of the data, while the ICH (and other FDA) guidances provide a framework for study design, including dose selection, duration of dosing, and safety endpoints. Together, these guidelines can have a significant impact on the formulation strategy over the course of nonclinical development.

One of the tenets of the GLP guidelines is characterization of the test article. The guidelines require an assessment of the "identity, strength, purity, and composition" of the test and control articles. In addition, the stability and uniformity/homogeneity as it is used on study (same formulation and storage conditions/time) must be established. In practice, most sponsors provide a single certificate of testing (CoT) establishing the identity, strength, and purity of the test and control articles, in addition to pre-study analysis of the stability (under appropriate storage conditions) and (if a suspension) homogeneity in the formulation. Stability of the TA in the formulation under the anticipated conditions of use is critical. For instance, if the TA is stable for less than 24 h in a given formulation, daily dose preparation will be required for the duration of the study, which is a significant resource drain and is generally not feasible for studies longer than several weeks. Formulation test article concentration and homogeneity are tested periodically throughout the study to ensure that animals are being dosed appropriately. All of these data are generated using GLP-validated assays and reported in the final report. Typically, the limit for test article content and homogeneity are ±10% of nominal. If a dose preparation fails to meet specifications, it is the job of the nonclinical toxicologist to assess the impact to the study based on the size and frequency of the deviation, available exposure data, and other available information. Formulations that cannot pass these assessments for content, stability, or homogeneity under conditions of the study should not be considered for use in repeat-dose GLP studies.

One of the most important questions to answer prior to GLP toxicology studies is how much TA will be required to run these studies. This number can vary greatly, depending mainly on the species, dose levels, and duration of the studies. For example, a potent kinase inhibitor with a top dose level of only 5–10 mg/kg in the rat will require less than 10 g for a 1-month study, while a study of similar duration in dogs or minipigs at top dose levels of 1 g/kg will require many kilograms of API. Table 4.3 demonstrates the approximate TA requirements for a variety of study lengths in rats and dogs at three different top dose levels—10, 100, and 1000 mg/kg. These estimates include a 25% overage required because of TA loss during dose preparation and transfer and required dose analysis testing. As you can see, the TA requirements can vary

Table 4.3 Example test article requirements for GLP general toxicology studies.[a]

Species	Rat			Dog		
Top dose	10 mg/kg	100 mg/kg	1000 mg/kg	10 mg/kg	100 mg/kg	1000 mg/kg
Approximate weekly volume per dose group	400 mL	400 mL	400 mL	3.5 L	3.5 L	3.5 L
Approximate weekly TA requirement (all groups)	1.6 g	16 g	160 g	14 g	140 g	1.4 kg
1-month study: approximate total TA requirement	6.4 g	64 g	640 g	56 g	560 g	5.6 kg
3-month study: approximate total TA requirement	20 g	200 g	2 kg	180 g	1.8 kg	18 kg
6-month study: approximate total TA requirement	42 g	420 g	4.2 kg	360 g	3.6 kg	36 kg

[a] Based on a typical GLP general toxicology study design and average body weight for an adult animal. Includes 25% additional material for loss in transfer and GLP dose verification testing assays.

greatly and can reach even higher for compounds tested for longer than 6 months or in larger species such as minipigs. As discussed previously, the selection of species and duration for toxicology studies requires an understanding of the cross-species metabolic profile and the clinical plan, making it very hard to estimate TA requirement of toxicology studies before this information is available.

4.4 Phase-Appropriate Nonclinical Formulation Strategy

The purpose of nonclinical formulations for orally administered compounds is to provide adequate exposure for nonclinical safety studies. As such, a flexible, phase-appropriate approach to nonclinical formulation strategy is most appropriate (Figure 4.1). As described in detail in this and the previous chapter, there are many study types employed in order to assess the safety of an NCE, and these will vary greatly in their duration, species used, and endpoints. Therefore, it is often not feasible to utilize the same formulation for the duration of a nonclinical program. Instead, a careful assessment of the goals of each study and the formulation requirements can provide an easier path through development.

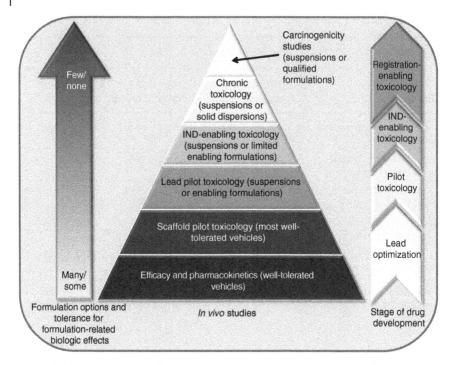

Figure 4.1 Phase-appropriate formulation strategy. As a program progresses from lead optimization to clinical and eventually registration-enabling studies, formulation for *in vivo* studies should evolve from grossly well-tolerated formulations with some biological effects to fully qualified GLP formulations with no such effects. Early lead optimization or pilot toxicology studies can utilize formulations with minor biological effects because of the limited safety endpoints and short duration of dosing. However, as the program matures, the formulation options become more limited, generally resulting in simple suspension formulations being used for chronic or carcinogenicity studies.

As described in the previous chapter, small molecule lead optimization studies often focus on *in vitro* potency, selectivity, and safety. As a program matures, *in vivo* efficacy, pharmacokinetics, and safety become critical. It is these initial *in vivo* studies that allow for the most options in terms of non-clinical formulations. *In vivo* pharmacokinetic studies assess exposure following both oral and intravenous in order to establish key pharmacokinetic parameters, including bioavailability. These are generally single-dose studies and can therefore be accomplished with solvent-based formulations, such as those containing DMSO, ethanol, PEG, Cremophor, and propylene glycol (Thackaberry *et al.*, 2014). While some of these solvents, such as PEG, may interfere with the bioanalysis of certain molecules, they are generally well tolerated after a single use and are extremely useful as platform screening formulations.

Initial *in vivo* efficacy studies may also employ a wide variety of formulations. The key criterion for these formulations is that they be grossly well tolerated while producing sufficient exposure for efficacy without interference with the primary efficacy endpoint(s). For example, a solvent-based formulation that causes mild liver toxicity would be acceptable in a short-term efficacy study for an NCE intended to treat extrahepatic diseases. In contrast, such a formulation may be unacceptable in assessing the efficacy of an NCE for nonalcoholic steatohepatitis (NASH). *In vivo* safety studies, often referred to as pilot toxicology studies, require a more careful assessment of formulation requirements. Early pilot toxicology studies often focus on de-risking particular on- or off-target liabilities, such as hepatotoxicity or GI toxicity, and are often performed with early non-optimized molecules. In these studies, a wide variety of formulations may be employed, as long as they do not interfere with the ability to de-risk the key safety liabilities.

In contrast, general repeat-dose pilot toxicology with the lead molecule or with a small set of molecules that may include the lead molecule requires a more "clean" formulation. These studies generally assess toxicity to multiple organ systems, including those often affected by high doses of vehicles, such as the liver and GI tract. In addition, these studies generally include detailed clinical and anatomic pathology endpoints, which may uncover formulation-related toxicities not observed in earlier pharmacokinetic and efficacy studies with the same formulation. Another key consideration at this point is the influence of formulation on toxicity. Even formulations that are inherently nontoxic at the doses utilized may affect the toxicity of an NCE due to their impact on exposure. When a formulation change increases exposure, for example, increasing AUC from 10 to 50 μM h at a fixed dose of 100 mg/kg, increased toxicity should be expected. Altering the rate of absorption, thereby modulating serum C_{max}, may also lead to a change in toxicity profile. Because of these potential effects on the toxicity profile, it is ideal to use a "GLP-friendly" formulation for pilot toxicity studies with potential lead molecules, if possible. This eliminates the need for bridging pharmacokinetic studies between pilot and GLP toxicology studies and provides more confidence in the predictability of the pilot toxicology studies. However, more enabling formulations may be used in these pilot toxicology studies in order to maximize exposure, if needed.

IND-enabling toxicology studies require formulations with fewer potential toxicological or tolerability liabilities. While the duration of these studies varies according to the planned clinical trial designs, such studies are often only 1-month long and may be as short as a single dose, which will impact the acceptable formulation options. As discussed later, the majority of approved molecules over the past several years utilize simple suspension formulations for these studies. However, more enabling options such as lipid, PEG, or cyclodextrin formulations may also be employed for these studies as required. IND-enabling GLP studies are generally run in two species, one rodent and one non-rodent. While the use of the same formulation in both species is convenient, it is not required, which allows for additional flexibility as the development program proceeds. For example, capsule dosing may be utilized in dogs while oral gavage of simple suspension formulations is used in rats.

As a molecule moves beyond phase 1 into larger clinical trials, additional toxicology studies are required (see Section 4.2). These include longer-term (chronic) general toxicology studies as well as specialized studies such as developmental and reproductive toxicology (DART) studies. At this stage, formulation options become even more restrictive due to the requirements of these studies. For example, while 100% lipid or greater than 20% cyclodextrin formulations may be acceptable for 1-month IND-enabling GLP toxicology studies, such formulations are generally not acceptable for chronic toxicology studies. At this stage, a more inert formulation, such as a solid dispersion, is often developed.

The final nonclinical safety studies for many NCEs are the carcinogenicity studies. Due to their duration and unique safety endpoints, these studies represent the highest bar for nonclinical formulations. Because of the resources and timelines involved in these studies, the FDA should be consulted on study design and dose selection prior to initiation, and the acceptability of the formulation should be included in this discussion. The specific issues and requirements for carcinogenicity study formulations are covered in more detail later.

4.4.1 Phase-Appropriate Formulation Options

Within the pharmaceutical industry, companies and laboratories often have preferred or default IND-enabling formulations, and their individual choices can vary greatly. However, a recent survey of successful IND/registration-enabling programs from 2000 to 2011 provides insights into the general trends within the industry (Thackaberry, 2013). The vast majority of these programs used simple suspension formulations, usually methylcellulose (or similar cellulose derivative) with or without a surfactant such as Tween 80. The advantages of these formulations are multifold; they are practically nontoxic, and high-concentration suspensions are much easier to produce than solutions for most modern NCEs. For compounds with relatively low solubility but adequate permeability, these formulations can often produce very good exposures simply by virtue of their high concentrations. For less soluble compounds, nanoparticle formulations may be an option. The increased surface area provided by these smaller particles may enhance GI absorption, resulting in significantly increased exposure using the same simple cellulose-based formulation.

For more soluble NCEs, a simple aqueous buffered solution may be adequate. However, such compounds are increasingly rare in the pharmaceutical industry, with poorly soluble/permeable compounds becoming more common. For these less soluble NCEs, enabling formulations may be required to produce sufficient exposure in the GLP safety studies. However, many of these vehicles carry with them toxicological or biologic effects in nonclinical species. As such, the choice of enabling formulation is often a balancing act between the enabling characteristics of the formulation and these unwanted effects. For shorter-term safety studies (1–3 months), several enabling options are available, such as HPβCD, corn oil, or PEG. However, for longer-term studies, particularly carcinogenicity studies, the enabling options can be very limited.

The simplest nonclinical formulations are simple solutions. When possible, water alone may be used. However, more commonly pH-adjusted buffers are used, typically acetate, citrate, or phosphate buffers with a pH between 3 and 10. These formulations are generally well tolerated when the concentration of the buffer is kept to a minimal level. A drawback is that they can only be used with highly soluble NCEs.

As mentioned earlier, suspension formulations are the most commonly used in the industry. These generally consist of an emulsifying/suspension agent such as methylcellulose (or derivative) or gum (generally acacia or tragacanth) at low concentration (generally 0.5–2%) with or without a surfactant (usually Tween or SDS). These formulations offer the advantage of being nontoxic while enabling high doses and high exposure with more permeable NCEs. The utility of the formulations is limited by the ability to create homogeneous suspensions or plateaus in exposure that may occur due to poor permeability or absorption.

Enabling formulations can take many forms and present both opportunities and challenges. As mentioned earlier, the use of these formulations is often a balancing act between exposure enhancement and tolerability. Lipid formulations (most commonly corn oil) can be used at lower-dose volumes or for shorter duration studies (≤1 month) in most species. The major disadvantage of lipid-based formulations is tolerability in large animals (diarrhea or emesis) and metabolic effects in rodents, which preclude their use on carcinogenicity studies. Tolerability of these formulations can also vary significantly based on the lipids used. For that reason, lipid mixture such as the so-called self-emulsifying drug delivery systems (SEDDS) cannot be assessed for tolerability based on their individual components but must be individually tested for tolerability in each species of interest. PEG 400 is another useful enabling formulation that can be utilized at low-dose volumes or at standard volumes for shorter-term studies. For many of these enabling formulations, especially lipid-based formulations, viscosity and palatability can be a concern. Highly viscous formulations should be avoided in repeat-dose toxicology studies. Besides being difficult to dose, such formulations are more likely to be aspirated into the lungs, leading to pulmonary toxicity. This is a particular concern on chronic toxicology studies. Palatability should also be considered—animals will struggle to avoid formulations that taste bad, often leading to dosing errors. In some cases, bad tasting formulations may cause excessive salivation in rodents, sometime prior to dosing as the animals anticipate the bad taste of test article.

Cyclodextrins are a very well-characterized vehicle option. They can boost exposure of poorly soluble NCEs with their unique encapsulation mechanism. Their use is limited by tolerability and toxicity issues that are dependent on the route of administration and species being assessed. Oral cyclodextrins are well tolerated in most species up to 10–20%, though they can cause elevations in transaminases in rodents at low doses (Thackaberry *et al.*, 2010), which may be unacceptable when trying to assess the safety of NCEs with potential liver toxicities. In larger animals, especially rabbits and monkeys, GI issues limit their use, with rabbits being particularly sensitive.

Amorphous dispersions and nanoparticulate formulations have become more common in the industry over the past several years. While these formulations may require significant resources to develop, they are generally well tolerated in nonclinical species. The limiting factor for both amorphous dispersions and nanoparticulate systems is the ratio of drug to polymer and the total polymer dose. Polymers used in these formulations, such as HPMCAS and PVP, are generally well tolerated but can produce GI effects if given in large doses. In these cases, the ratio of polymer to NCE should be kept as low as possible in order to limit the polymer dose and reduce viscosity.

4.5 Methods of Test Article Administration

By far the most common method of NCE administration in nonclinical studies is intragastric gavage. This is generally done via the oral route but may also be performed nasally in larger species (particularly nonhuman primates). Dosing solutions/suspensions are withdrawn in an appropriate-sized syringe (generally 1–100 mL) and administered directly to the stomach via a gavage needle (for rodents) or a catheter/intubation tube (for larger animals). Technical information on typical gavage systems across the nonclinical species are given in Table 4.4. Reflux and aspiration can be issues, particularly with low-pH or solvent formulations that can potentially damage the esophagus or lungs. To prevent this, vehicle or saline flushes (typically 5 mL) are often used in larger animals. Even with the use of a saline flush and careful test article administration, some of the formulation is likely to be deposited in the esophagus from the outside of the needle/tube, especially with viscous formulations. Therefore, the pH of the dosing formulation should generally be kept within the 3–10 range. Dose volumes can easily be calculated from the most recent body weight, which is an advantage for fast-growing rodent species, and allows for accurate dosing.

While oral gavage is used for the majority of oral nonclinical studies, other methods of administration may be employed. These alternative methods are often utilized for convenience on chronic studies but may also be used for specialized studies or to address toxicity or tolerability concerns.

Table 4.4 Considerations for oral gavage across standard nonclinical species.

Species	Gavage volumes (mL/kg)[a]	Gavage setup[b]	Typical dose volume[c]
Mouse	5–10	20 gauge, 5.5 cm ball- or tear-tipped stainless steel intubation needle	200 µL
Rat	2–10	14–16 gauge, 9 cm ball- or tear-tipped stainless steel intubation needle	3.75 mL
Rabbit	2–10	35–40 cm No. 12–18 Fr. catheter	20 mL
Minipig	1–10	50–75 cm No. 16–24 Fr. catheter	150 mL
Dog	1–5	50–53 cm No. 15–32 Fr. catheter	100 mL
Monkey	1–5	35–47 cm No. 12–15 Fr. catheter (oral)	20 mL
		33–40 cm No. 8–9 Fr. catheter (nasogastric)	

[a] Recommended dose volumes for repeat-dose toxicology studies based on Diehl *et al.* (2001).
[b] Catheter use may vary by laboratory and size/age of the animal. Syringe selection should be based on the dose volume.
[c] Based on a 5 mL/kg dose volume in an average adult male animal.

Dietary administration is commonly used in chronic rodent studies, particularly carcinogenicity studies and the associated dose-ranging studies. The main advantage of dietary administration is a reduction in animal handling/stress and elimination of daily gavage procedures, which can be a significant resource savings. Dietary administration may also reduce GI toxicity and aid in absorption. Test article is added to the feed in order to achieve the desired dose level based on the known food consumption for the given species, strain, and age of the animal. Remaining feed must then be weighed in order to estimate total feed intake and by extension test article intake based on current body weight. While dietary administration is commonly used, it does come with several disadvantages. Pilot studies should be conducted to determine if the test article is stable in feed, if the feed remains palatable to the test species, and if exposure is sufficient. High variability may be encountered due to feed spillage, and effects on food consumption or body weight secondary to the toxicity of the test article may have profound dose-related effects on exposure. This highlights the importance of toxicokinetic analysis to characterize exposure across all dose groups. Additionally, the nominal dose levels (mg/kg) may change with age and will generally be different between males and females, since food consumption changes with age and female rodents tend to eat more per gram of body weight than males. Despite these issues, dietary administration can be extremely useful on chronic rodent studies with well-defined NCEs that have little or no impact on body weight or food consumption.

Capsule dosing is also commonly used in chronic studies, generally for dogs. Again, convenience is a major driver, as it is generally much easier and less stressful to the animal to administer capsules in dogs. In addition, these powder-in-capsule formulations are often identical to those used in phase 1 clinical studies, which allows a more direct comparison from dog to human, and can facilitate specialized nonclinical studies, such as food-effect studies. Generally, a weeks' worth of capsules are filled based on weekly body weight measurements, though more frequent fills may be required. While stability of test article is not an issue with capsule dosing, pilot studies are still needed to confirm acceptable exposure. Additionally, the maximal dose is limited by the size of capsules available, though multiple capsules may be given.

Finally, there may be some situations in which a non-oral formulation would be required for a drug intended for oral administration in the clinic. These include investigative toxicology studies addressing C_{max}-related toxicities, abuse liability studies, or reformulations for parenteral administration. Most commonly intravenous administration is utilized in these studies. In these cases, the formulation need not resemble the oral formulation or use compendial excipients, as long as the formulations meet the criteria set in Table 4.1.

4.6 Formulation Tolerability Across Species and Study Designs

There are no FDA or ICH guidelines governing the selection of nonclinical formulations for NCEs intended for oral administration. For non-oral drugs, the guidances suggest that the clinical formulation is used nonclinically and often require bridging pharmacokinetic studies when the formulation changes over the course of a development program. However, for oral molecules, the sponsor is left to their own devices. This is because of the numerous compendial oral excipients available and universal acceptance of exposure as the driving factor behind safety and efficacy of pharmaceutical agents. As such, nonclinical studies may utilize any available formulation, so long as they produce sufficient exposure to cover the anticipated clinical exposure, and conform to the other principles laid out in Table 4.1. While in theory this provides limitless options for nonclinical formulations, in practice, the options are often quite limited. In addition, the available options change significantly based on the study design and duration. Other factors that have a significant impact include the species, sex, age/developmental stage of the test system, safety endpoints assessed, therapeutic indication, and expected toxicity of the test article.

4.6.1 Vehicle Dose

One of the most common misunderstandings between the formulator and toxicologist is the subject of vehicle/excipient dose. From a toxicological perspective, the dose–response relationship is fundamental. "The dose makes the poison" is as true today as when Paracelsus coined the term in the sixteenth century. In nonclinical studies utilizing large doses in small animals, the dose volume of a formulation invariably exceeds clinical use. For example, a typical nonclinical dose volume of 5 mL/kg would equal a volume of 400 mL for an average human. Using the largest common capsule size (000), such a dose would require 295 pills without any excipients! Since the dose volumes are so much lower in humans, percent of excipients/vehicles is not a useful way of comparing toxicity between nonclinical and clinical formulations.

For example, ethanol is used at up to 70% in orally administered syrups without issues. However, at the larger dose volumes of nonclinical studies, 70% of ethanol is not tolerated in repeat-dose studies. This is because toxicity is driven by dose rather than percentage of an ingredient in a formulation. If ethanol were dosed at 70% at 5 mL/kg/day to dogs, the daily dose to the animal would be 35 mL, which is approximately equal to two shots of 80 proof spirits per day for an animal about one sixth the size of the average adult human. Therefore, clinical use of an excipient at a given percentage should not be considered supportive of nonclinical use at a similar level. When considering the acceptable levels of a vehicle, the total daily dose (expressed as mg/kg/day) should always be taken into account, which requires an understanding of the percentage of the vehicle, the dose volume, and the frequency of dosing (QD vs. BID).

4.6.2 Study Design

The tolerability of a vehicle or formulation will depend largely on the specific study design and toxicity endpoints assessed. The most obvious effect is the duration of the study. The majority of common nonclinical vehicles are well tolerated after a single dose, but as the number of doses increases, the tolerated vehicle dose decreases. One example of this is PEG 400, which has an oral LD_{50} in rats > 30 g/kg. PEG 400 has been reported to be tolerated at 10 g/kg for up to 1 month of daily dosing (Gad *et al.*, 2006). However, in a 3-month study, renal toxicity was noted above 1 g/kg in this species (Hermansky *et al.*, 1995). Dosing duration also affects tolerability endpoints that are not often associated with frank toxicity. For instance lipid formulations such as corn oil are generally well tolerated in short-term studies in dogs and monkeys (<1 week). However, with longer-term studies, diarrhea often becomes an issue. While diarrhea may be considered "non-adverse" in some instances, this can lead to serious health issues in animals over time, including body weight loss and malnutrition. While corn oil would not be considered toxic at these doses, the GI effects often prevent its use in chronic large animals.

While duration of dosing predictably lowers the tolerable dose level of most formulations, the specialized endpoints in other safety studies may produce less predictable formulation-related effects. For the majority of NCEs, some form of developmental or reproductive toxicity testing is required prior to large-scale studies in women of childbearing potential. For non-oncology oral small molecule drugs, four studies are generally performed (Table 4.2): fertility (in rat), teratogenicity (in rats and rabbits), and perinatal–postnatal development (in rats). For drugs intended to treat children, juvenile toxicity studies may also be performed (generally in rats). Of these studies, the most sensitive is the teratogenicity study, which assesses the effect of the drug on the organogenesis window of development. Some solvents, most notably ethanol, are known teratogens and should be avoided in these studies, even at low doses. The use of rabbits for teratogenicity studies, which was required in the wake of the thalidomide incident, significantly affects formulation selection. This is because the rabbit is not typically used for other toxicology studies (so previous experience with the formulation is unlikely), and they tend to be quite sensitive to GI issues, possibly due to their coprophagic nature. For example, 500 mg/kg of HPβCD in pregnant rabbit has been shown to cause GI issues leading to body weight loss and spontaneous abortions (Enright *et al.*, 2010). While this vehicle is not teratogenic, it is not useful as a vehicle in rabbit teratogenicity studies for this reason.

For safety pharmacology studies, care should be taken to avoid formulations that may interfere with the sensitive endpoints involved. For cardiovascular safety pharmacology, intravenous administration of cyclodextrins should be avoided, since they alter hemodynamic parameters (Rosseels *et al.*, 2013). Cyclodextrins should also be avoided in *in vitro* hERG assays, where they can mask drug effects (Mikhail *et al.*, 2007). In CNS safety pharmacology studies, high doses of solvents should be avoided due to their potential to cause narcosis.

The most limiting studies in terms of acceptable formulations are carcinogenicity studies. These studies are designed to assess carcinogenic risk following a lifetime exposure (2 years) in mice and rats. Recently, transgenic mouse models have been developed that allow for shorter duration studies (6 months) in mice, but the rat carcinogenicity assay still requires 2 years of dosing. For carcinogenicity assessment, the safety endpoint is tumors or preneoplastic/neoplastic lesions, the incidence of which are assessed compared to concurrent controls and historical controls using the same study paradigm. Since many tumor types are rare even when induced by exposure to a carcinogen, these studies require large animal numbers (50–75/sex/group) and in depth statistical analysis to identify carcinogenic risk. As such, any formulation that alters the baseline incidence of tumor formation is likely to make interpretation of the study nearly impossible, since such incidence would not be in line with historical controls. The most well-documented example of this is lipids. A National Toxicology Program (NTP, 1994) study investigated the effects of corn oil, safflower oil, and tricaprylin in 2-year mouse and rat carcinogenicity studies and found dramatic dose-related differences in tumor incidence in both species. While these lipids are not considered carcinogenic, the chronic exposure to high doses of lipids clearly alters the background incidence of several tumor types, compromising study interpretation, and therefore should be avoided. Other examples of vehicles that may alter tumor incidence in carcinogenicity studies include ethanol (known tumor promoter), vitamin E TPGS (antioxidant), and HPβCD, which may induce pancreatic tumors in rats (Gould and Scott, 2005).

One final study type worth noting is abuse liability testing. These studies are generally only performed for NCEs with known abuse liabilities based on mechanism of action within the CNS, though they may also be performed for CNS-active drug with a poorly defined mechanism of action. For such NCEs, abuse liability testing usually includes self-administration and drug discrimination studies, which assess the tendency of animals to seek out the drug and similarity of effect to known classes of abused drugs, respectively. In both of these study types, an intravenous formulation is generally required, since only IV administration will result in the rapid pharmacologic effect required for these study designs.

4.6.3 Species-Specific Tolerability Issues

The species used in nonclinical toxicology studies can impact the tolerability observed. Rodents tolerate most vehicle components better than the typical non-rodent species (dog and monkey), though they may be more sensitive to some vehicles, such as cyclodextrins, which have been shown to cause elevations in liver transaminases when administered in rats but not in other species. More commonly, the non-rodent species is more sensitive. This is often due to the difference in GI function, with dogs and monkeys much more prone to diarrhea and also subject to emesis, which is not observed in rodents. The higher absolute volumes of formulations given to these larger animals' species may also play a role. While diarrhea and emesis are not normally considered significant toxicities, they can both be dose limiting. Emesis often reduces drug exposure, particularly when it occurs immediately post dose. This can also lead to damage to the esophagus and a reduction in body weight if food intake is affected. Diarrhea can be even more serious in these species, especially monkeys. Consistent diarrhea can cause significant weight loss and is not compatible with continued dosing at extended period of time. Both emesis and diarrhea are significant animal welfare concerns and should be considered an unacceptable formulation-related reaction in repeat-dose toxicology studies. Another species-specific tolerability issue mentioned earlier is that of oral cyclodextrins, which cause GI issues in rabbits and should be avoided in this species.

While this chapter is focused mainly on oral administration, other routes of administration may be employed for oral NCEs and can lead to tolerability issues. Probably the best example of this is the use of nonionic surfactants in dogs. Dogs tolerate low doses of oral nonionic surfactants, a class that included Tween, Pluronic, and Cremophor. However, even at low doses, these surfactants cause an anaphylactoid reaction when given intravenously in this species. These reactions can be severe and even lead to death of the animal; therefore nonionic surfactants should not be used in intravenous formulations intended for use in dogs. The toxicity profile of many vehicles changes significantly based on the route of exposure, so oral tolerability should not be used as a surrogate for toxicity data via other routes. Another example of this is the cyclodextrins, which are generally well tolerated when given orally at doses of up to 1000 mg/kg but which cause significant renal damage at these doses when administered intravenously.

4.7 The Relationship between Clinical and Nonclinical Formulations

As mentioned earlier, oral clinical and nonclinical formulations are generally distinct. The FDA regulates the use of excipients in clinical formulations and maintains the inactive ingredient database in order to clearly establish acceptable limits of use. In contrast, there are no guidances on the composition of nonclinical formulations. This underscores a fundamental difference between clinical and nonclinical ingredients. Inactive ingredients in clinical formulations are demonstrated to be "safe" through nonclinical and clinical testing in support of a drug—generally the first drug to use an inactive ingredient or push the percentage use of the ingredient above previously accepted levels. In contrast, oral nonclinical studies are very rarely designed to assess the safety of the formulation. Within pharmaceutical development, the term "safety" refers to human safety. Therefore, for a nonclinical formulation that is not intended to be used in the clinic (or is already in the inactive ingredient database), there is no need to assess the safety of the formulation—only the safety of the NCE is assessed. This will be the case for the vast majority of oral nonclinical programs. The only exceptions (discussed in the next few paragraphs) are when novel excipients or significantly higher levels of previously approved excipients are intended for use in the clinic.

The lack of guidances on nonclinical formulation selection allows the sponsor to select any formulation that meets the criteria outlined in Table 4.1. The relationship of nonclinical safety findings to clinical safety is then established by comparing exposure between nonclinical and clinical studies. The relationship between exposure and safety is critical and has been well established with oral drugs. In this context, exposure is expressed in terms of both AUC and C_{max} at steady state, and exposure multiples can be calculated as described previously. The lack of guidances in this area allow for numerous nonclinical formulations options, including the use of a wide range of common vehicles, novel vehicles, and non-gavage administration (capsules, dietary administration). Another tactic that has been successfully employed is the coadministration of exposure-boosting compounds. Generally, these are cytochrome P450 inhibitors, such as ritonavir, which inhibit the metabolism of the NCE, enhancing exposure to the parent molecule significantly.

Due to the impact of physical form on pharmacokinetics (and, by extension, safety profile), the same crystalline form and (if applicable) salt form intended for use in the clinical should be utilized on all safety studies. Unlike nonclinical vehicles, which are generally not intended to be used clinically, nonclinical safety studies assess the toxicity of both the NCE and the counterion of therapeutic salts. Therefore, the salt form should be tested as early as possible in the development program, and if late salt changes are made, bridging pharmacokinetic and/or toxicology studies may be required.

4.7.1 Novel Excipients

The use of a novel excipient in the clinic requires that the safety of the novel excipient be established nonclinically. This can be done in parallel with an NCE (i.e., in the same studies) or separately. The FDA has published a guidance for the nonclinical safety testing of novel excipients (FDA, 2005), which is largely the same battery of tests required for a new (non-oncology) drug (Table 4.2). The parallel development track, in which the novel excipient is tested alongside an NCE in the same studies, is the most efficient, even if a second vehicle control group is required to control for effects of the novel excipient. While stand-alone studies may provide a more robust toxicological assessment by exploring a dose range and potentially dosing above the anticipated clinic use to identify potential toxicities, such studies are rarely performed due to the high cost of the full nonclinical testing battery. Current regulatory practices do not allow for completely separate development program for novel excipients; novel excipient safety data must be submitted to the FDA along with safety data for a NCE. Unfortunately, this regulatory paradigm limits the qualification of novel excipients, since even the most well-tolerated and effective novel excipients will fail to be codified in the FDA inactive ingredient database if the associated NCEs fails to be approved for clinical use in the United States. More commonly, available excipients are utilized at levels higher than previously approved in an attempt to enhance a clinical formulation. In this case, new nonclinical data may be generated, though it may not be required if data on the safety of the excipient is available in the literature.

While the current regulatory guidances do not detail the studies required to bring a novel excipient into early clinical trials (instead focusing on studies required for registration), common practice is to follow the ICH M3 guidance in which nonclinical data equaling for exceeding the duration of intended clinical use is sufficient for entry into the clinic. Industry groups such as the International Pharmaceutical Excipient Council (IPEC) (Velagaleti *et al.*, 2009) and the IQ Consortium are working toward a more streamlined and well-defined process for the development of novel excipients, and hopefully this process will become more clear in the near future.

4.8 Conclusions

Nonclinical formulation selection is an ongoing process over the course of the development of an NCE. In order to maximize exposure while minimizing formulation-related risk across multiple study designs and species, a phase-appropriate formulation approach is warranted, allowing for more platform-enabling formulations early in the development and replacing these as the molecule advances with inert formulations capable of supporting chronic and carcinogenicity testing. Because the nonclinical safety program is heavily dependent on the intended patient population and expected treatment duration, study design, species selection, and doses required, it is critical for the formulation scientist to work closely with the nonclinical toxicologist in order to assure that these required studies are adequately supported with formulations meeting the criteria outlined in Table 4.1 of this chapter. This flexible approach is key to maximizing the probability of success across an oral small molecule development program.

References

Code of Federal Regulations (CFR) Title 21 (1978) Chapter 1: Food and Drug Administration, Department of Health and Human Services; Subchapter A: General; Part 58: Good Laboratory Practices. Available at: http://www.ecfr.gov/cgi-bin/text-idx?SID=733225d718e1d16156ab52bfdca494c1&mc=true&tpl=/ecfrbrowse/Title21/21cfr58_main_02.tpl (accessed October 7, 2016).

Diehl, KH, Hull, R, Morton, D, Pfister, R, Rabemampianina, Y, Smith, D, Vidal, JM, van de Vorstenbosch, C. (2001) A good practice guide to the administration of substances and removal of blood, including routes and volumes. *J. Appl. Toxicol.* 21:15–23.

Enright, BP, McIntyre, BS, Thackaberry, EA, Treinen, KA, Kopytek, SJ. (2010) Assessment of hydroxypropyl methylcellulose, propylene glycol, polysorbate 80, and hydroxypropyl-beta-cyclodextrin for use in developmental and reproductive toxicology studies. *Birth Defects Res. B Dev. Reprod. Toxicol.* 89(6):504–516.

Food and Drug Administration (2005) Guidance for Industry: Nonclinical Studies for the Safety Evaluation of Pharmaceutical Excipients. Available at: http://www.fda.gov/downloads/Drugs/GuidanceComplianceRegulatoryInformation/Guidances/UCM079250.pdf (accessed on October 7, 2016).

Gad, S, Cassidy, CD, Aubert, N, Spainhour, B, Robbe, H. (2006) Nonclinical vehicle use in studies by multiple routes in multiple species. *Int. J. Toxicol.* 25(6):499–521.

Gould, S, Scott, RC. (2005) 2-Hydroxypropyl-beta-cyclodextrin (HP-beta-CD): a toxicology review. *Food Chem. Toxicol.* 43:1451–1459.

Hermansky, SJ, Neptun, DA, Loughran, KA, Leung, HW. (1995) Effects of polyethylene glycol 400 (PEG-400) following 13 weeks of gavage treatment in Fischer-344 rats. *Food Chem. Toxicol.* 33(2):139–149.

International Conference on Harmonization of Technical Requirements for Registration of Pharmaceuticals for Human Use (ICH) (1995) Guideline on the Need for Carcinogenicity Studies of Pharmaceuticals (S1A). Available at: http://www.ich.org/fileadmin/Public_Web_Site/ICH_Products/Guidelines/Safety/S1A/Step4/S1A_Guideline.pdf (accessed on October 7, 2016).

International Conference on Harmonization of Technical Requirements for Registration of Pharmaceuticals for Human Use (ICH) (2000a) Safety Pharmacology Studies for Human Pharmaceuticals (S7A). Available at: http://www.ich.org/fileadmin/Public_Web_Site/ICH_Products/Guidelines/Safety/S7A/Step4/S7A_Guideline.pdf (accessed on October 7, 2016).

International Conference on Harmonization of Technical Requirements for Registration of Pharmaceuticals for Human Use (ICH) (2000b) Detection of Toxicity to Reproduction for Medicinal Products and Toxicity to Male Fertility (S5). Available at: http://www.ich.org/fileadmin/Public_Web_Site/ICH_Products/Guidelines/Safety/S5/Step4/S5_R2__Guideline.pdf (accessed on October 7, 2016).

International Conference on Harmonization of Technical Requirements for Registration of Pharmaceuticals for Human Use (ICH) (2006) Impurities in New Drug Substances (Q3A). Available at: http://www.ich.org/fileadmin/Public_Web_Site/ICH_Products/Guidelines/Quality/Q3A_R2/Step4/Q3A_R2__Guideline.pdf (accessed on October 7, 2016).

International Conference on Harmonization of Technical Requirements for Registration of Pharmaceuticals for Human Use (ICH) (2009a) Guidance for Nonclinical Safety Studies for the Conduct of Human Clinical Trials and Marketing Authorization for Pharmaceuticals (M3). Available at: http://www.ich.org/fileadmin/Public_Web_Site/ICH_Products/Guidelines/Multidisciplinary/M3_R2/Step4/M3_R2__Guideline.pdf (accessed on October 7, 2016).

International Conference on Harmonization of Technical Requirements for Registration of Pharmaceuticals for Human Use (ICH) (2009b) Nonclinical Evaluation for Anticancer Pharmaceuticals (S9). Available at: http://www.ich.org/fileadmin/Public_Web_Site/ICH_Products/Guidelines/Safety/S9/Step4/S9_Step4_Guideline.pdf (accessed on October 7, 2016).

International Conference on Harmonization of Technical Requirements for Registration of Pharmaceuticals for Human Use (ICH) (2011) Guidance on

Genotoxicity Testing and Data Interpretation for Pharmaceuticals Intended for Human Use (S2). Available at: http://www.ich.org/fileadmin/Public_Web_Site/ICH_Products/Guidelines/Safety/S2_R1/Step4/S2R1_Step4.pdf (accessed on October 7, 2016).

International Conference on Harmonization of Technical Requirements for Registration of Pharmaceuticals for Human Use (ICH) (2013) Photosafety Evaluation of Pharmaceuticals (S10). Available at: http://www.ich.org/fileadmin/Public_Web_Site/ICH_Products/Guidelines/Safety/S10/S10_Step_4.pdf (accessed on October 7, 2016).

Mikhail, A, Fischer, C, Patel, A, Long, MA, Limberis, JT, Martin, RL, Cox, BF, Gintant, GA, Su, Z. (2007) Hydroxypropyl β-cyclodextrins: a misleading vehicle for the *in vitro* hERG current assay. *J. Cardiovasc. Pharmacol.* 49:269–274.

National Toxicology Program (NTP) (1994) Comparative Toxicology Studies of Corn Oil, Safflower Oil, and Tricaprylin (CASRNs 8001-30-7, 8001-23-8, and 538-23-8) in Male F344/N Rats as Vehicles for Gavage. National Toxicology Program Technical Report Series 426:311, US Dept of Health and Human Services, Washington, DC.

Rosseels, MLA, Delaumois, AG, Hanon, E, Guillaume, PJP, Martin, FDC, van den Dobbelsteen, DJ. (2013) Hydroxypropyl-β-cyclodextrin impacts renal and systemic hemodynamics in the anesthetized dog. *Regul. Toxicol. Pharmacol.* 67:351–359.

Thackaberry, EA. (2013) Vehicle selection for oral safety studies. *Expert Opin. Drug Metab. Toxicol.* 9(12):1635–1646.

Thackaberry, EA, Kopytek, S, Sherratt, P, Trouba, K, McIntyre, B. (2010) Comprehensive investigation of hydroxypropyl methylcellulose, propylene glycol, polysorbate 80, and hydroxypropyl-beta-cyclodextrin for use in general toxicology studies. *Toxicol. Sci.* 117:485–492.

Thackaberry, EA, Wang, X, Schweiger, M, Messick, K, Valle, N, Dean, B, Sambrone, A, Bowman, T, Xie, M. (2014) Solvent-based formulations for intravenous mouse pharmacokinetic studies: tolerability and recommended solvent dose limits. *Xenobiotica* 44:235–241.

Velagaleti, R, DeMerlis, C, Brock, W, Osterberg, R, Goldring, J. (2009) Regulatory update: The IPEC novel excipient safety evaluation procedure. *Pharm. Technol.* 33(11).

5

Planning the First Clinical Trials with Clinical Manufacturing Organization (CMO)

Elizabeth Kwong[1] and Caroline Mcgregor[2]

[1] Kwong Eureka Solutions, Montreal, Quebec, Canada
[2] Merck Research Laboratories, Kenilworth, NJ, USA

Oral Formulation Roadmap from Early Drug Discovery to Development,
First Edition. Edited by Elizabeth Kwong.
© 2017 John Wiley & Sons, Inc. Published 2017 by John Wiley & Sons, Inc.

5.1 Reasons for Outsourcing

Despite today's challenging environment, pharmaceutical industry players continue to strive to deliver scientific excellence, meet unmet medical needs, and drive state-of-the-art innovation. At the same time, the industry and its stakeholders are cognizant that the current model is still "broken." To fix this situation, one particular analysis of R&D productivity (Paul *et al.*, 2010) has pointed toward the reduction of costs and cycle time as a prime opportunity, as well as bringing compound attrition earlier during lead optimization and preclinical evaluation and before the first-in-human stage of clinical development. In other words, the "preclinical stage" of investment is now becoming the new "phase IIb proof-of-concept study" for key "go/no-go" decisions. Those authors suggest that this paradigm shift will increase the overall probability of technical success in later stage phase II and III studies. Smarter management of resources is required, and this has shifted thinking from the more traditional approaches of internalizing efforts to more cost-effective alternatives such as external outsourcing. Outsourcing if done properly can fit seamlessly into the business design of the pharmaceutical industry.

In fact, outsourcing of manufacturing activities for drug products (DP) has become increasingly prevalent in the industry over the last 10–15 years, and this trend is expected to continue. In addition to the reasons stated previously, the ongoing divestment of these internal capabilities by big pharma and the rise of virtual start-up companies that depend fully on contract research organizations (CROs) and contract manufacturing organizations (CMOs) have contributed to the current trend (Levy, 2014; Van Arnum, 2013).

Most companies that outsource usually try to take advantage of economies of scale and find ways to leverage either work load or capabilities for favorable pricing from the outsourcing organization. This chapter covers first clinical trials, and therefore the focus will be on CMOs and not CROs.

Sponsors have the choice of determining what to outsource and why; however, to simplify the discussion, the chapter will also assume that the entirety of the clinical trial material (CTM) manufacturing activities are being outsourced.

In the outsourcing industry, it had been common knowledge that cost efficiency coupled with high quality work is most attractive to the client. Secondary considerations include the timelines that encompass scheduling, meeting agreed timelines, executing overall turnaround, and responsiveness in communication with scientific personnel, business development, and project managers. In addition, CMO decision-makers have clearly articulated the importance of service quality, setting priorities, and the ability to deliver rapid, effective problem solving in their services. Other important attributes of the CMOs include adequate resources, logistical considerations such as location of the laboratories relative to the good manufacturing practices (GMP) manufacturing facility, and the client such that shipping of materials to and from this facility will not be impacted by specific regulations in a given country, quality of the report, flexibility, and agility to manage potential study changes as well as history of prior relationship.

5.2 Considerations for Outsourcing

Rather than simple transactional or fee-for-service outsourcing, preferred-provider relationships are now being practiced regularly by the pharmaceutical industry to provide deeper, more collaborative, and hence enhanced strategic partnerships with the CMOs. Both big pharma and smaller biopharmaceutical companies are using this same model. Some recent examples include the Pfizer collaboration with Icon and Parexel (Miller, 2014), where Pfizer has retained scientific ownership of the clinical development process and is maintaining oversight and quality standards relating to patient safety and regulatory compliance. By using this preferred-provider model, Pfizer has claimed that they have managed to reduce the number of partners they were dealing with from 18 to 20 plus while still obtaining quality work and cost savings. For start-up companies who lack infrastructure, brick, and mortar laboratories, the preferred-partner approach allows them to maintain familiarity with the CMO's process and leverage the expertise of their scientists while managing the timelines, quality, and cost appropriately. The goal is synergy where the scientists from the sponsor and the CMO work side by side as colleagues. From a recent experience, it was also deemed important to establish the right partnership with a CMO that supports the same phase clinical studies with the client. For example, if a CMO only have experience with late stage or commercial products, their standard operating procedure (SOP) will only cover activities in the development of late stage products. The types of support expected in late stage or commercialization are more complex and may prolong timelines. From this current experience it was observed that these CMOs are usually less flexible as you would encounter during early phase. This small oversight can contribute to some difficulty in determining the types of activities required when handling an early phase project.

5.2.1 CRO/CMO Perspective

A recent paper by Yarger (2014) summarized some of the requirements a CMO must meet in order to work synergistically with the client:

1) Fully comprehend what the company needs, and be able to show examples of how their experience and expertise can address these needs.
2) Have world-class capabilities combined with expertise and experience.
3) Have scientists and technical experts who are able to think scientifically and creatively while keeping things simple.

4) Offer additional/alternative interpretation of data that they help to generate.
5) Have a proven track record of completing projects on time and on budget.
6) Be willing to admit when they lack the requisite expertise for a specific initiative.
7) The use of project managers to organize team activities such as agenda, minutes, follow-up assignments, and timeline is critical for execution of a project.
8) Share with clients details of any regulatory inspections and number of client audits to demonstrate true capabilities to manufacture GMP supplies and manage partnerships.
9) Establish a network of companies offering related services, for example, analytical microbial testing lab, qualified person (QP) for clients wishing to perform studies in Europe, specialty testing labs such as preformulation studies, salt selection, and polymorph screening.

5.2.2 Sponsor's Perspective

There are also criteria that the sponsor must follow in order for this partnership to work:

1) The sponsor must openly provide clear project goals and objectives in their request for proposal (RFP) and in ongoing communications. This will eliminate the potential for future friction between competing partner needs:
   a) This clarity will be important for the CMO to match its services and expertise. High scientific standards and robust knowledge should be a requirement for any CMO in this synergistic relationship. When a sponsor determines that the CMO has insufficient expertise, the sponsor will avoid using that CMO in the future and will not recommend them to other sponsors who are pursuing the same services with this CMO. When necessary, the sponsor may terminate an unsuccessful relationship before the work is completed. This can be very risky for the sponsor due to how critical the drug products will be to the timeline to initiate the first clinical trials.
   b) The CMO can contribute to the refinement of the project goals and objectives.
   c) While the CMO may streamline the details of the timeline and schedule, the sponsor and innovator always remains in control of the overall project, including areas such as research methodology and data analysis and interpretation. In most cases these should be discussed and approved by the sponsor. (Note: It is important that the data interpretation around technological aspects is shared and agreed with the CMO.)
   d) It is also important that the sponsor be transparent with the CMO and share proprietary information with the CMO, subject to the appropriate "confidentiality disclosure agreement (CDA)." This will allow CMO to understand

the science behind the contracted work and the reason for the scientific questions underpinning the work. A high level of trust is required beyond the standard confidentiality disclosure agreement. Open discussions of any gaps in understanding or experience involved should be encouraged on both sides.

e) When CMO is engaged due to their specialty capabilities, for example, spray drying or nano-formulation approaches, it is important that the sponsor examines thoroughly what needs to be done and ensures that readily accepted technologies are used rather than sophisticated new analytics that may be difficult to explain in an Investigational New Drug (IND) application. The intellectual property (IP) position should be clarified up front and in writing before contracting research to the CMO. Most sponsors will need to retain their IP and will have difficulty in engaging with a CMO that has clearly built its business around a unique patent position.

f) The best CMO is not always the least expensive. Outsourcing has to be cost effective in the long term and efficient, and the CMO must provide the right amount of data to support the project. However, the service provider also has to be mindful of financial constraints such that services are affordable and delivered on time.

g) When entering into a synergistic partnership, it is very important to have an on-site visit to enable more efficient and informed data analysis and decision-making. On-site visits should be viewed as a means to help both parties understand the details behind the procedures being used. The ability to interact directly with technicians allows opportunity to highlight any unforeseen aspect of the data generated by the investigators. It is not unusual that with the accumulated knowledge of the scientists from the sponsor and service provider, such discussions may yield additional discoveries that could otherwise be lost during virtual meetings, remote teleconferences, and e-mails. Regular on-site visits and subsequent observations can facilitate timely adjustments to the research process, as well as minimizing or clarifying any confusion or misalignment of expectations.

h) A project lead or single point of contact from the sponsor can help lessen miscommunication between partners.

i) The use of experienced early phase drug products consultants can also facilitate the negotiation of contracts as well as introductions to other experts with specialized knowledge in those cases where deeper understanding of the formulation is required.

j) Clear timeline expectations had to be charted out from the onset. The timeline can be accelerated by using "do it right the first time" strategies as part of the development process.

5.3 Timing for CMO Selection

Efforts to select the right CMO partner for the drug products is critical to the timeline of the project as this will be the last piece of information required for inclusion in the regulatory filing. Ideally, the timing of the selection will be aligned with that for the active pharmaceutical ingredient (API). The API batch size will depend on the design of the clinical trials and subsequent clinical supply needs (i.e., batch size and dose strengths) and should be incorporated into the API planning to ensure that only one batch needs to be manufactured to support phase I, thus maintaining costs and eliminating frequent updates to the filing. Selection of the drug products CMO should take into consideration the complementarity of capabilities of the API CMO to ensure that all of the required information for the filing is generated in a timely manner while minimizing duplication of effort and information. For example, details of the preformulation report should be laid out clearly. Solubility information for the API should be complete such that it includes evaluation of the current form's solubility in organic solvents, aqueous systems including biorelevant media, and buffers to assess pH-dependent solubility. Assignment of responsibility and ownership should be clear such that the information is delivered, but there is no duplication by either the drug substance (DS) or drug products CMO to limit confusion as to which number is reliable or should be reported in the regulatory documentation.

Additionally, the location of each CMO should be considered such that shipping of API to the drug products facility will not itself cause any challenges. For example, it will be important to identify up front the complexities in shipping across borders, for example, if the API CMO is in Canada but the drug products CMO is in the United States, despite the close proximity, customs requirements can be complicated for moving materials across the border.

5.4 Pre-CMO Selection Background Information/ Preparation

5.4.1 Detailed List of "Must-Have" Information Before Establishing a Partnership

A secure IT platform and data repository enables communication and sharing of information between partners without compromising internal networks, security, or IP:

1) Early draft of the clinical plans to understand:
    a) Size of phase I study and hence batch size
    b) Timeline to file
    c) Formulation strategy, for example, extemporaneous or on-site formulation preparation, API in capsule, and enabled formulation

    d) Dose strengths required

    e) Packaging requirements, for example, bottles (with number of counts), blister pack, or bulk packaging

    f) Potential requirements for a double-blind, placebo-controlled clinical study or a comparator product

2) Where to file, for example, the United States, Europe, or Canada

3) Readiness for a quality assurance (QA) audit of the CMO facility

4) Availability of representative API for development and GMP manufacturing

5.4.2 Predevelopment Information

As the candidate selection activities are ongoing, it is very important that an adequate assessment of proper developability of the molecules is conducted as outlined in references such as the commentary from *Journal of Pharmaceutical Sciences* (Kwong, 2015). The physicochemical characteristics of the selected form will be critical information to be shared with any potential CMO after signing a CDA. This will allow a proposal to be developed up front with the appropriate formulation strategy for the drug products. Physical properties including hygroscopicity and chemical properties such as photostability are especially important considerations to be included in the proposal to ensure that the cost, for example, of packaging is realistically assessed and captured. Knowing hygroscopicity of the materials to be used can help inform on the type of excipients to consider, processing parameters to be used to avoid issues, and the need for packaging with desiccants or other barrier-providing materials such that they can be planned into the statement of work (SOW). Such basic and general information can be included to narrow down the scope of the proposal. Additionally, these details can be shared again during the launch of the project.

    Moisture sorption analysis is usually performed as part of the early preformulation evaluation to characterize the profile of moisture uptake of various crystalline salt/neutral forms and confirm acceptable stability across the relative humidity (RH) range. Callahan *et al.* (1982) established four classes of hygroscopicity: very hygroscopic, moderately hygroscopic, slightly hygroscopic, and nonhygroscopic. These are based on both the amount and rate of water uptake over various RH ranges. Later, the European Pharmacopoeia adopted a simpler approach where they measured the amount of water uptake at 25°C and 80% RH and classified solids as outlined in Table 5.1. However, the utility of both of these classifications is limited as they provide little understanding of why water is taken up, and hence they will not guide handling, processing, or packaging of the API (Ahlneck and Zografi, 1990; Reutzel-Edens and Newman, 2006; Zografi, 1988). More recently Newman *et al.* (2008) developed a flowchart that provides a basis for using water sorption/desorption data obtained under typical RH conditions used during processing of pharmaceutical solids.

Table 5.1 Hygroscopicity classification.

Classification	Criteria per Callahan *et al.* (1982)	Criteria per EU Pharmacopoeia[a]
Nonhygroscopic	Class I: essentially no moisture ↑ below 90% RH; <20% ↑ in moisture content above 90% RH in 1 week	0–0.012% (w/w)
Slightly hygroscopic	Class II: essentially no moisture ↑ below 80% RH; <40% (w/w) ↑ in moisture content above 80% RH in 1 week	0.2–2% (w/w)
Moderately hygroscopic	Class III: moisture content does not ↑ >5% (w/w) below 60% RH; <50% (w/w) ↑ in moisture content above 80% RH in 1 week	2–15% (w/w)
Very hygroscopic	Class IV: moisture content will ↑ as low as 40–50% RH; >20% (w/w) ↑ in moisture content above 90% RH in 1 week	>15% (w/w)

[a] Percent water uptake at 25°C/80% RH.

They claimed that the flowchart will provide a systematic assessment of hygroscopicity for the API.

Murikipudi *et al.* (2013) published a more efficient throughput method for hygroscopic classification using water sorption analysis that they compared to the EU and Callahan classification. They concluded that using dynamic vapor sorption (DVS) had the advantage of providing a complete assessment of the sorption and desorption profile of the sample while conserving both time and material. This approach provides a better correlation to the conventional method reported by Callahan *et al.* (1982) with no overlap of categories while also providing information on potential solid-state transformation under various moisture conditions.

Powder flow is also a key physical property of the active ingredient in the pharmaceutical manufacturing process. It is essential to measure this prior to capsule filling or tablet manufacturing. There are several compendial methods to measure powder flow: (i) bulk and tap density, (ii) Hausner ratio, (iii) Carr's compressibility index, and (iv) measurement of angle of repose. It is noted in the USP that no one simple powder flow method will adequately or completely characterize the wide range of flow properties. It has been suggested that an appropriate strategy may be to use multiple standardized test methods to characterize the various aspects of powder flow (Powder Flow USP, 2012). Table 5.2 describes the classification of powder flow based on Hausner ratio, compressibility index, and angle of repose as described in the USP.

Significant photoinstability can also be a liability during processing, particularly if exposure to ambient light is sufficient to degrade the active ingredient.

Table 5.2 Evaluation of flow properties.[a]

Flow properties	Angle of repose (degrees)	Compressibility index (%)	Hausner ratio
Excellent	25–30	≤10	1.00–1.11
Good	31–35	11–15	1.12–1.18
Fair—aid not needed	36–40	16–20	1.19–1.25
Passable—may hang up	41–45	21–25	1.26–1.34
Poor—must agitate, vibrate	46–55	26–31	1.35–1.45
Very poor	56–65	32–37	1.46–1.59
Very, very poor	>66	>38	>1.60

[a] USP General Chapters: <1174> Powder Flow.

The use of lighting controls such as a sodium lamp can mitigate the risk, but this may be a capability that the CMO does not have.

In addition to any chemical stability and physical characterization information such as solubility in biorelevant media, exposure in preclinical species used to support toxicology studies will also be important information to be shared since this will serve as a guide to determine the complexity (i.e., simple blend in capsule vs. needs for enabling formulations such as spray-dried solid dispersions) of the dosage form to be developed for the clinical trial.

5.5 Selection of CMO

5.5.1 Process of CMO Selection

5.5.1.1 Request for Proposal

Following signing of a CDA, RFP or request for quotation (RFQ) for drug products development should be prepared while negotiating with the API CMO. The RFP should describe development of the prototype formulation, analytical development, GMP manufacturing activities, shipping, and storage of inventory including the timeline of individual activities and corresponding cost. A single vendor approach where manufacturing and analytical laboratories are located within the same organization is always beneficial with regard to timeline. The typical timeline from development of the prototype formulation to release of the GMP CTM can range from 6 to 9 months depending on the complexity of the formulation and communication between the partners. Including two or more projects within the same RFP may not be practical as multiple projects may compete for the same resources and thus impact timeline. Detailed discussion of placing two or more projects into one CMO should

Table 5.3 Example of cost breakdown.

Item	Activities	Cost
1	Health and safety evaluation	
2	Project management	
3	Analytical methods (product release, in-process testing methods, microbial testing, qualification, release testing) • Cleaning validation • API ID testing for internal CoA • Product methods development and qualification	
4	Preformulation (flow properties, excipient compatibility study, solubility in excipients, etc.)	
5	Formulation development/selection (prototypes of lowest and highest dose strength)	
6	Engineering/demonstration run	
7	Stability studies (demo/GMP batch)	
8	GMP clinical trial manufacturing (preparedness, batch records—manufacture/packaging/labeling)	
9	Storage and shipping	

be initiated as soon as appropriate developability assessment information are available for the candidates in order to initiate development.

Requests usually contain sufficient information to allow the CMO business manager to provide an appropriate quote/proposal/SOW. The CMO proposal should be customized to your development needs, and thus, initiating with a standard templated proposal can prolong the negotiation process and delay alignment and approval. Some CMOs may provide an RFQ template to help the customer provide the information that they will need to deliver an accurate proposal. Proposals are usually provided by the CMO after signing their CDA as this will allow them to then share their company's approach to providing the required services. Approval of the proposal will need to include review by the technical scientists involved in the project as well as the chief financial officer or procurement manager (for small start-up pharma) who will be responsible for negotiation of the terms of payment/cost (Table 5.3), and so on.

1) Contents of the RFP:
   a) Introduction of the sponsor:
      i) Information on the projects (generic/high level) and approach to development
      ii) Expectations for GMP CMO contract organization
      iii) How proposal will be evaluated

2) Contact information: to whom proposal should be addressed
3) Proposal or SOW:
    a) Type of formulation: API in capsule, formulated capsule, enabled formulation such as hot melt or spray-dried formulation, and so on
    b) Phase of study, that is, phase I with resupply scheduling, for example, x months
    c) Dose strengths with or without placebo
    d) Batch size
    e) Statement of work:
        i)   Communication strategy needs and plan
        ii)   Raw material purchase, analytical test components
        iii) Method transfer from API CMO
        iv)   Development of prototype formulations including supply for biopharmaceutical evaluation on preclinical species
        v)   Feasibility study to determine scalability and to supply the prototype stability study and analytical method development
        vi)   Stability protocol requirements for the prototype formulation or demonstration batch and GMP manufacture
        vii) GMP preparedness, manufacturing, packaging, labeling, and release
        viii) Storage and shipping
        ix)   Timelines for the project
4) Specific work plan:
    a) Detailed formulation development activities (assumes preliminary work has been done based on exposure in preclinical species)
    b) Detailed development of analytical methods including types of methods required, internalization of methods, qualification protocol of methods, release methods, stability study protocol, and specification for release and stability
    c) Formulation screening approach, number of prototypes, stability studies, and feasibility/scalability
    d) Batch records (manufacturing, packaging/labeling)
    e) Release data (timing of certificate of analysis (CoA))
    f) Facility audit requirement
    g) Proposed timeline
    h) Communication needs and plan, for example, teleconference, face to face, minutes, agenda, and so on
5) Additional information:
    a) CV of development formulators and analytical chemists
    b) Letters of reference from present and former clients of CMO
    c) Customer compliance audit
    d) Regulatory agency inspection
    e) Capabilities and number of successful manufacture for different types of formulation and batch size manufacturing experience

    f) Equipment list

    g) Facility layout

    h) Risk assessment process

    i) Health and safety evaluation of the API (occupational health categorization and compound handling/MSDS)

    j) Detailed breakdown of pricing

5.5.1.2 Soliciting Proposals

The quality of the proposals received will be dependent on the RFP that the sponsor initially shared with the CMOs. It is very important that the RFP is carefully constructed to ensure that the desired type of proposal is received. How the CMOs conduct themselves during the RFP process often reflects how they will conduct business under contract. This will allow the sponsor to benchmark the CMO for alertness, attentiveness, timeliness, and problem solving capabilities. The proposal should include a Technical Agreement that contains desired specifications and a Quality Agreement that should include the following:

1) Provision for frequent visits
2) Oversight during manufacturing of the sponsor's products
3) Requirements for prior approval of any changes to the facility, equipment, materials, and/or components
4) Notification of any out-of-specification (OOS) test results or deviations and stability failure, including stability chamber failures
5) List of documents for sponsor approvals including master batch records, deviations, and/or investigations

5.5.1.3 Comparison Criteria

To complete the RFP process, it will be very beneficial to compare the performance attributes of each CMO to narrow down selection of the preferred partner(s) (Miseta, 2015). A detailed approach to evaluating the partnership with the CMO is a "must" and should be practiced consistently. As the sponsor compares the different CMOs, it is important to always ask if a particular CMO can be the long-term partner you are expecting?

    Table 5.4 is an example of the comparison criteria that can be used to ensure that the selected CMO meets the sponsor's expectations.

    Each attribute should be graded according to the importance of the attribute to the product that will be developed and delivered. Examples of detailed requirements of each attributes enumerated in the table are as follows:

1) **Capabilities** will include the types of formulation the CMO can make and scale up in the GMP facility. Not all CMOs are equal and therefore may not have the same equipment that can provide the quality of drug products needed for clinical trials. Or the CMO may not have the necessary infrastructure to

Table 5.4 Comparison of performance attributes of CMOs.

Attributes	CMO 1	CMO 2	CMO 3
Capabilities			
Scale-up (size)			
Management team			
Turnover rate of employees			
Delivery histories (timeliness)			
References			
Regulatory inspections (citations)			
Site overview			
Cost			

support, for example, generating the master batch record for such a formulation. Capability assessment should also include the analytical laboratory in both development GMP areas. The instruments being used should be current and have the appropriate operational qualification/performance qualification (OQ/PQ). The analysts must be knowledgeable about the different analytical methods used to determine quality of the product. Examples of the qualification protocol should be reviewed to determine the thoroughness of the laboratory. However, it is also important that the CMO demonstrate an understanding of the phase-related activities such that they are not using late phase requirements when working on an early phase product. At the end of the day, the sponsor needs to prove that the CMO has the ability to consistently manufacture compliant, safe, and effective quality products. Also determine in this category whether there are any language barriers or lack of responsiveness to e-mail or phone that can exacerbate potential contractual issues.

2) **Scale-up** activities in the GMP facility must be assessed such that timing of the manufacturing will not cause delays because, for example, the CMO has to manufacture several batches of the same dose strength or that this can lead to ballooning of the cost of the SOW. It is also important that the scale-up activities are not too far from the development laboratory since this may contribute to very complicated technology transfers.

3) **Management team** should be knowledgeable of the regulatory guidances and competent in managing any test failure investigations, deviations, and out of specifications. The team must understand the stage-appropriate activities to ensure that work is not overdone. The quality of products/results should be scrutinized carefully before sharing with the sponsor. Timeliness of the reporting of results also has to be a priority with the team.

Yearly quality checks of SOPs should be performed. Training records should be current. These are just some of the strengths of the team that the sponsor needs to address with the CMO up front during the comparison.

4) **Employee turnover rate** information can also be important in assessing potential partnerships with the CMO. High turnover rate can be a reflection of the quality of the management and the CMO itself. On the floor discussions with employees during the walk-through can in some cases provide hints as to the quality of the scientists that the client will eventually rely on to make their product. Most service providers have the same technologies in-house. The scientists who are operating the technologies will be the difference the sponsor may be looking for.

5) **Historical delivery information** should also be assessed. The experience of the CMO at a particular stage of development gleaned from batch delivery information and development history can be an indication of their capabilities to deliver the types of product that will be needed by the sponsor. Timeliness of the delivery can also be evaluated if the information is available. In some cases it can be indirectly determined from references. Gantt charts for the project should be included in the SOW to allow for alignment with the sponsor's desired timeline and project plan.

6) **References** from the CMO can also help determine the suitability of the CMO to deliver the desired product. The best reference will be someone you can trust to provide the unambiguous truth about their experience with the CMO. The most important performance attributes the sponsor must be able to determine from the references are the quality of the products produced previously and the timeliness of the delivery. Most often the CMO will be selected based on a reference before the evaluation. These references can be gathered from conferences, past experience with the CMO, and word of mouth. However, since projects will differ from one customer to another, it is very important that a thorough evaluation is still performed and aligned with each individual customer's need.

7) **Regulatory inspections (citations)** experience is an important factor in assessing the CMO. Pre-approval inspection (PAI) records from the Food and Drug Administration (FDA) should be reviewed. A waived PAI can provide an indication of the quality of the CMO that might ultimately affect your product. Records of corrected regulatory issues, gaps, and nonconformities should also be reviewed. In addition to the regulatory inspection records, deviations and corrective action preventive action (CAPA) closures must also be reviewed.

8) **Site visits** while not the most important comparison criteria are a key component. The visit, in addition to the QA vendor audit visit that will be addressed in the next section, will be the sponsor's true measure of the

CMO's capabilities. The site visit will allow not only an assessment of the knowledge of the group but also the layout of the facilities and whether this flows logically such that the risk of cross-contamination of products, from other sponsors, is avoided. Most importantly, the ability of the CMO to provide the facilities, personnel, and systems to deliver services that are compliant with current good manufacturing practices (cGMPs) can be assessed. It is preferred that the physical plant has all of the unit operations for the desired process and product in one location. This is especially important for "virtual" companies and start-ups that do not have manufacturing capabilities of their own. It is very important that the sponsor be prepared with specific questions that they want to be addressed during the site visit. The business development representatives from the CMO are generally very knowledgeable, but the technical staff should be able to provide some specific insights to the approach they would use to address problems that might arise.

Even after selection of the CMO, the sponsor should conduct regular site visits for the purpose of ongoing assessment of manufacturing quality. As part of these frequent site visits, oversight of manufacturing processes and review of quality test data and records should be conducted.

9) **Cost** assessments must be compared thoroughly since the SOW or proposal may not be broken down in the same way and may not include the same activities across each individual CMO. This is where the quality of RFP and the specific details become very important in ensuring that the sponsor can easily and appropriately compare across the CMOs. This can be a challenging task since it can often depend on the approach of the business manager representing the CMO. In some circumstances it can be helpful to schedule a teleconference to review details and build an in-depth understanding of the proposal. The comparison must be done by someone with appropriate understanding of the process of development through GMP manufacture. This may pose a challenge for start-up companies that usually will not have a development scientist on staff. This is where a consultant can be helpful, but it will be very important to select a consultant with direct experience of running early phase trials and who actually understands this space. Many of the consultants might be more familiar with late stage development. The key point is to have a phase-appropriate strategy and not over-engineer early phase development.

10) In **follow-up**, to close out efforts around soliciting proposals, it will be very important to inform the CMOs as to the outcome of the proposal reviews. For the few CMOs that criteria were met, a follow-up visit and audit should be scheduled.

5.5.1.4   Other Documents: MSA and Quality Service Agreement (QSA) (Quality Performance)

The conditions that govern the relationship between the service provider and sponsor may be described in a master service agreement (MSA). The following outline for the MSA attempts to address all of the issues surrounding quality, IP, confidentiality, performance, timelines, SOW, reports, payments, audits, and terms of the agreement including termination. This section represents the highlight of the MSA and will depend on the sponsor.

The MSA must include, at a minimum, specific information related to the two partners (CMO and sponsor) with a specified effective date. A definitions section clarifying the contents and types of service should be included. The services should include the proposal, performance, change orders, timelines, and project reports. The payment section should specify fees and expenses with due dates for payment. In this section, it will be helpful to clarify how to handle disputed amounts. Also clarify the types of audits that the sponsor would like to conduct during this agreement period to be clear on expectations from both sides. In most MSA, a quality performance section is also included that outlines the expected quality of the drug products based on agreed specifications and includes the quality of the analytical methods. Payment or failure to deliver fees for instances where there is a quality failure should be agreed in advance and outlined in the MSA. It should be also clarified that the contract agreement will also be pending QA audit.

Agreement on how to protect each party's IP should be clarified in this document with regard to ownership, disclosure and assignment, and license grants to company. Lastly, confidentiality of information including exceptions, authorized disclosure, publication, return of confidential information, and acknowledgment of injunction relief must be specified in the agreement.

5.5.1.5   QA Vendor Audit

Before selecting the preferred CMO partner, a QA audit by the sponsor is one of the most important confirmations of whether the CMO being considered is maintaining appropriate quality standards. Consequently, choosing the QA auditor should be considered very carefully. The auditor should be someone who has tenure in the industry to be able to understand good quality systems and who has a strong working knowledge of the process. This individual should have extensive experience balanced with appropriate people skills to manage the interaction.

According to Howe and Winberry (2010), there are four stages of effective vendor audits: audit planning and preparation, conducting the audit, the audit report, and audit responses and follow-up. Most of these audits can be completed within 2–3 days.

5.5.1.6   Audit Planning and Preparation

Planning should be tailored to the purpose of the audit visit, that is, selection of CMO and routine audit of an active CMO, or related to a specific cause such as the CMO being out of compliance. The audit team should include the auditor accompanied by representatives from the sponsor, most likely with complimentary expertise. No more than two people should be added, and this depends on the scope of the audit. An agenda for the audit should be shared with the CMO that includes the following:

1) Tour of facilities involved in the manufacturing and release of the drug products including material receipt, sampling and storage, dispensing and staging areas, sample retention and reference storage, manufacturing, raw material, in-process, release and stability testing laboratories, stability chambers, packaging and labeling, label and packaging storage, drug products storage and shipping areas, and handling of potent compounds.
2) Procedure/document/record reviews and discussions will cover but are not limited to site master file, validation master plan, quality manual, contamination control/product changeover and cleaning verification, method development/method transfer/method qualification and validation, OOS/out-of-trend reporting, risk management program, change control, deviation reporting, investigations, corrective/preventive action program, management reviews, training program, internal audits, batch record review and product disposition, complaints, product recall/retrieval, supplier management program, and annual report processes.
3) Regulatory inspection history including examples of other client audits.
4) Selected installation qualification (IQ)/OP/PQ of equipment used.
5) Water system validation.
6) Environmental monitoring program/results and trending.
7) Vendor qualification and technical agreements including other partners used in the testing and release of product, for example, microbial testing.

5.5.1.7 Conducting the Audit

The lead auditor will be responsible for recording the observations raised and will also author the report. That individual should clarify the roles and responsibilities before the audit and manage them during the audit. A clear agenda with timing must be followed to ensure that each item is covered. Any critical issues must be recorded and followed up. It is always beneficial to resolve any conflicts by discussion, by clarifying points, and by effective communication. Observations must be clearly raised during the audit. This will allow the CMO to investigate if this is truly an issue and also determine if it can be closed out before the end of the audit if easily resolved.

During the introductory meeting, it will be important for the sponsor and the auditor to understand the business model of the CMO, for example, sustainability, track record of profitability, and growth. Although this may be a sensitive issue, it will be very important for this audit to determine the CMO's intention for sustainability including selling or merging with another existing business. It will also be important to meet the oversight team for the project that will be contracted to build strong partnerships that can be beneficial in managing tight timelines when the project is underway. The facility tour will provide an impression of the strengths and weaknesses of the company's commitment to quality. It will be crucial to take the time to observe the condition and status of rooms and equipment, as this is often an indication of status of work and quality.

5.5.1.8 Closeout Meeting

Before the closeout meeting, the auditing team should have a private meeting to review overall feedback. Categorize the observations by regulatory, quality, facility setup, and so on.

It is recommended that the closeout meeting should start with any positive observations noted during the audit. All observations should be clearly expressed and response of the host be taken into consideration. This meeting should highlight the timelines for the audit report and for response related to any critical observations and the complete response to the audit report once it has been received by the CMO.

5.5.1.9 Audit Report

Every effort should be made to write and approve the report within 10 working days of audit completion. The report should be structured and written clearly with unambiguous language. Typically the report should include an introduction of the auditor and purpose of the audit, summary, observation including criticality, and conclusions:

1) **Introduction** should include reference to the standards audited against a brief description of the facility being audited, attendees, and copy of the agenda.
2) **Summary** should have a high level summary of the audit findings (e.g., number and categorization of observations) and should refer to the documentation reviewed during the audit.
3) **The observations** should be structured by type of observation and criticality. It is common to include recommendations for actions to be taken in response to observations.
4) **Concluding remarks** should clearly state the timeline requirements for responses and the conclusions of the auditors with reference to the purpose of the audit.

5.5.1.10 Response and Follow-Up

It will be very important that not only the responses are received but also that there is follow-up to confirm that the agreed actions have been completed. This is usually followed by an additional audit to truly close out the responses.

The lead auditor should be in regular contact with the CMO to establish a timeline for follow-up on outstanding observations not completed. In some cases the final audit reports can be tracked in the CAPA system to ensure that proper closeout of the observations occurs. It will be very important to track the responses and follow-up carefully to ensure that the sponsor is completely satisfied with the outcome and status.

5.5.2 Criteria

According to a recent survey (Outsourcing Survey, 2014), sponsors identified a series of criteria and key areas of focus when selecting a CMO. Figure 5.1 illustrates these selection criteria from the sponsors listed in order of importance.

It is vital to properly identify a good CMO partner particularly when critical work and the timeline of the project are dependent on the success of the formulation. Remember that no two CMOs are alike; they will have different strengths and weaknesses. Key elements in the selection criteria should include quality, timeliness, flexibility, technical expertise, size of the facility, size of the company, confidentiality, facility compatibility, capacity, customer service, and price. Some criteria to consider are as follows:

1) *Reputation:* Seek out CMOs with a strong reputation with regard to IP integrity, especially when considering non-Western vendors and CMOs. Make sure to understand the country's approach to protecting IP assets. Some countries offer no protection, and corporate fraud is rampant. IP litigation is expensive and time consuming.

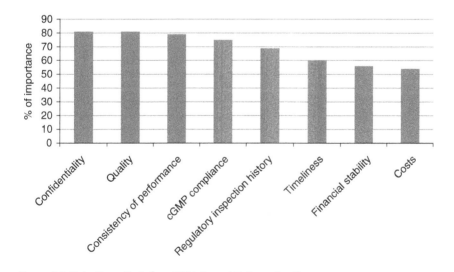

Figure 5.1 Selection criteria from 2014 Annual Outsourcing Survey.

2) *Expertise:* CMOs that provide strong technical expertise are always critical since this are key to managing tight timelines, efficiency, and thorough appreciation of the SOW.

3) *Communication:* Understand the barriers to communication such as language and time zones. Determine if adequate communication skills and attention to detail/thoroughness exist. Understand their openness, honesty, and ability to provide up-front, candid discussion.

4) *Company practices:* Understand how the company operates with regard to hierarchy, decision-making, report writing, analytical data turnaround, deviation reporting, safety evaluation of API, handling of potent compounds if needed, management of confidentiality related to the product being manufactured, inventory storage, shipping, and so on. This is usually addressed during a site visit where the sponsor should ask direct questions. It will also be important to understand how the company manages long cycle time activities within their timelines and balances relative to other business.

5) *Matching culture:* Consider the CMO's reputation and reliability. Are the culture and personalities of the employees very different to those of the sponsor?

6) *Scheduling flexibility:* It is very important to also understand the *other business and* competing projects in the CMOs portfolio to ensure flexibility of scheduling of the sponsor's GMP manufacturing activities.

7) By leveraging a combination of these comparison and selection criteria evaluated during vendor selection, the sponsor should be able to finalize the signing of the contract for the planned work. Signing this contract will initiate and solidify the "partnership" such that information transfer can begin and the project can move to implementation.

5.6 Final Thoughts

Figure 5.2 summarizes the critical steps involved in selecting the right CMO partnerships. The details of each of these steps have been described in the preceding sections of this chapter. Careful attentions to these details are a crucial part of identifying the right vendors for effective execution of the project plan to support early first-in human clinical studies with the right focus on collaboration.

In summary, it is very important to note that mutual respect had to be established between CMOs and sponsors. Starting a relationship with the CMO had to be considered carefully. Issues that can create problems with the CMO relationship involved misaligned expectations between the CMO and the sponsor and can usually be showstoppers in the relationship and can have disastrous ramifications for both parties. According to data released in

Primary key	Product	Batch no.	D[v,0.1] μm	D[v,0.5] μm	D[v,0.9] μm	D[4,3] μm
2008-08-13 17:20:29.6620 1257 H PF-04191834	E010008197	0.50	1.78	4.02	2.06
2008-08-13 17:22:10.9580 1257 H PF-04191834	E010008197	0.48	1.74	4.04	2.05
2008-08-13 17:24:34.8170 1257 H PF-04191834	E010008197	0.50	1.75	3.97	2.03

Figure 2.9 Particle size reduction of PF-04191834 after wet-bead milling with bench-top equipment.

Oral Formulation Roadmap from Early Drug Discovery to Development,
First Edition. Edited by Elizabeth Kwong.
© 2017 John Wiley & Sons, Inc. Published 2017 by John Wiley & Sons, Inc.

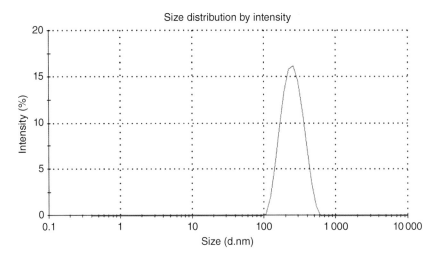

Figure 2.9 (*Continued*)

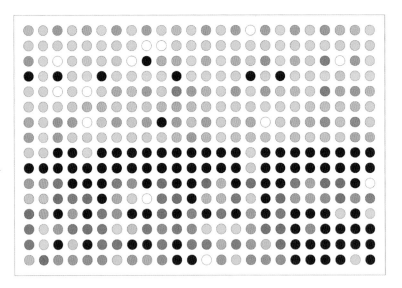

Figure 6.2 Visual heat map of particle sizes following acoustic mixing of a sample of naproxen with varying polymer and surfactant excipients over time, where green indicates small nanoparticles, red indicates larger nanoparticles, and black indicates lack of stable nanoparticles.

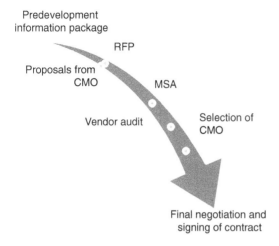

Figure 5.2 CMO selection steps.

the 8th Annual Report and Survey of Biopharmaceutical Manufacturing Capacity and Production (Bioplan Associates, Inc., 2011), the top 10 mistakes sponsors make with their CMO are as follows:

1) CMO attempted to manage costs by doing fewer runs leading to failures because poor process development results in insufficient process controls being established.
2) Sponsors don't build in sufficient time for the project.
3) Sponsors expect the CMO to resolve the most difficult scientific or technical problems.
4) Sponsors don't plan their technology transfer process appropriately.
5) Sponsors don't communicate with the CMO effectively.
6) Sponsors don't recognize the potential variability in process development.
7) Sponsors don't appreciate the differences between small-scale and full-scale manufacturing.
8) Sponsors don't understand their role in the regulatory submissions.
9) Sponsors expect the CMO to share with them their proprietary, internal process development, or manufacturing expertise.
10) Sponsors just hand off a project without planning for ongoing interactions.

Additionally, some of the CMO's typical pain points were also highlighted (Bioplan Associates, Inc., 2011), and they include the following:

1) Sponsors don't understand why GMP manufacturing costs are so high.
2) Sponsors have unrealistic timelines for approving and reviewing GMP documentation.

3) Sponsors refuse to devote adequate time to process development. The CMO can be pushed into GMP manufacturing prematurely that results in the need for process development work at scale in the clean rooms that will be very expensive.
4) Sponsors have no in-house QA/QC expertise and expect CMO to make regulatory decisions regarding their product.
5) Sponsors do not appreciate the importance of a comprehensive contract that specifies the roles and responsibilities of each party, especially when dealing with issues.

According to the data published in this report, the relationships between clients and the CMO are improving and shifting. For clients, quality, IP, and regulatory compliance will continue to be placed among the most critical of CMO attributes.

Abbreviations

API	active pharmaceutical ingredient
CAPA	corrective action preventive action
CDA	Confidentiality disclosure agreement
cGMP	current good manufacturing practice
CMO	contract manufacturing organization
CoA	certificate of analysis
CRO	contract research organization
CTM	clinical trial materials
DVS	dynamic vapor sorption
FDA	Food and Drug Administration
IQ	installation qualification
MSA	master service agreement
OQ	operational qualification
PAI	pre-approval inspection
PQ	performance qualification
QA	quality assurance
QSA	quality service agreement
RFP	request for proposal
RFQ	request for quotation
SOW	statement of work

References

C. Ahlneck, G. Zografi. The molecular basis for moisture effects on the physical and chemical stabilization of drugs in the solid state. *Int J Pharm* **62** (1990): 87–95.

Bioplan Associates, Inc. *8th Annual Report & Survey of Biopharmaceutical Manufacturing Capacity and Production* (A study of Biotherapeutic Developers and Contract Manufacturing Organization). Bioplan Associates, Inc., Rockville, 2011, pp. 1–22.

J.C. Callahan, G.W. Cleary, M. Elefant, G. Kaplan, T. Kensler, R.A. Nash. Equilibrium moisture content of pharmaceutical excipients. *Drug Dev Ind Pharm* **8** (1982): 355–369.

D. Howe, L. Winberry. Effective auditing of CMOs. *BioPharm Int Suppl* (2010). Available at www.biopharminternational.com (accessed November 2, 2016).

E. Kwong. Advancing drug discovery: a pharmaceutics perspective. *J Pharm Sci* **104** (2015): 865–871.

S.G. Levy. Effective outsourcing of small molecule chemistry R&D and API manufacturing for emerging pharmaceutical companies—a stepwise approach to risk management. *Pharm Outsourcing* **15**(1) (2014): 14–21.

J. Miller. Positive outlook for outsourcing. Pharm Technology supplement issue 2 (2014). Available at www.pharmtech.com (accessed November 2, 2016).

Ed Miseta. How to qualify a CMO's capabilities and benchmark its performance. Life Science Leader (Feb 26, 2015), pp. 1–4.

V. Murikipudi, P. Gupta, V. Sihorkar. Efficient throughput method for hygroscopicity classification of active and inactive pharmaceutical ingredients by water vapor sorption analysis. *Pharm Dev Technol* **18**:2 (2013): 348–358.

A.W. Newman, S.M. Reutzel-Edens, G. Zogfrafi. Characterization of the "hygroscopic" properties of active pharmaceutical ingredients. *J Pharm Sci* **97**:3 (2008): 1047–1059.

Outsourcing Survey. 2014 Annual Outsourcing Survey. Contract Pharma. Available at: http://www.contractpharma.com/contents/view_outsourcing-survey/ 2014-05-11/2014-annualoutsourcing-survey, 2014 (accessed October 4, 2016).

S.M. Paul, D.S. Mytelka, C.T. Dunwiddie, C.C. Persinger, B.H. Munos, S.R. Lindborg, A.L. Schacht. How to improve R&D productivity: the pharmaceutical industry's grand challenge. *Nat Rev Drug Discov* **9** (2010): 203–214.

Powder Flow USP. US Pharmacopeia General Chapter <1174> Powder Flow, 2012.

S.M. Reutzel-Edens, A.W. Newman. The physical characterization of hygroscopicity in pharmaceutical solids. In: Hilfiker, R. (Ed.). *Polymorphism in the Pharmaceutical Industry*, pp. 235–258, Wiley-VCH, Weinheim, 2006.

P. Van Arnum. A shifting landscape for the global API market. *Pharm Technol* **2013**(4) (2013): 516–520.

J. Yarger. Working with CRO's: lessons for small biotechs. Contract Pharma, June 3, 2014.

G. Zografi. States of water associated with solids: a review. *Drug Dev Ind Pharm* **14** (1988): 1905–1926.

6

Formulation Strategies for High Dose Toxicology Studies: Case Studies

Dennis H. Leung[1], Pierre Daublain[2], Mengwei Hu[3] and Kung-I Feng[3]

[1] Small Molecule Pharmaceutical Sciences, Genentech, Inc., South San Francisco, CA, USA
[2] Discovery Pharmaceutical Sciences, Merck Research Laboratories, Boston, MA, USA
[3] Discovery Pharmaceutical Sciences, Merck Research Laboratories, Kenilworth, NJ, USA

6.1 Introduction

The discovery and development of new drugs has become increasingly challenging. The number of new drug candidates that encounter failure in clinical development has increased and the number of new drugs reaching the market has decreased in recent years (Munos, 2009; PhRMA, 2012; Scannell *et al.*, 2012). One factor contributing to the high attrition rate in drug development is unforeseen toxicity in the clinic (Kola and Landis, 2004). As a result, it has become increasingly important to conduct a more comprehensive assessment of the potential safety profile and adverse effects of new drug candidates in the preclinical discovery space prior to significant investment in clinical development. This may

Oral Formulation Roadmap from Early Drug Discovery to Development,
First Edition. Edited by Elizabeth Kwong.

help improve the safety profile of new drug candidates that progress into clinical trials and reduce the overall rate of attrition in drug development.

There are a number of early *in vitro* toxicology screens that can be implemented in the discovery space to assess potential safety liabilities such as genetic toxicity, off-target binding, and enzyme inhibition, among others (Bass *et al.*, 2009; Blomme and Will, 2016; Bowes *et al.*, 2012; Kramer *et al.*, 2007). However, in order to fully assess the safety profile of a new drug candidate, *in vivo* preclinical safety toxicology studies are often required to evaluate the effects of administering the drug or active pharmaceutical ingredient (API) to a live animal, typically a rodent and non-rodent species. These studies require administration of the drug at a wide range of doses and systemic exposure concentrations for an extended duration of time (Bass *et al.*, 2009). In order to facilitate these studies, a strong collaboration is required between discovery safety assessment groups and formulation scientists to ensure that sufficient drug is delivered reproducibly at each dose level to support the study design (Higgins *et al.*, 2012; Kwong *et al.*, 2011; Li and Zhao, 2007; Maas *et al.*, 2007; Shah and Agnihotri, 2011; Venkatesh and Lipper, 2000).

For these preclinical safety toxicology studies, generally the route of administration should match the route intended in the clinic, although there may be exceptions depending upon the desired effect to be monitored (e.g., intravenous delivery of a drug candidate to determine its impact on cardiovascular function). High drug exposure margins are required at the top dose levels ideally using the same formulation. This enables determination of the maximum tolerated dose (MTD), which is the dose that results in an exposure that does not produce an unacceptable toxicity, and the no-observed-adverse-effect level (NOAEL), which is the overall drug exposure that produces no adverse effects. This information is critical for helping to determine the therapeutic index or safety margin of a drug candidate. Consistent and reproducible exposures for each dose level are required, and near-dose linearity is highly desired in order to clearly distinguish any effects observed at different doses.

As a result, these safety studies typically entail formulations that involve high dose and drug loading, often significantly higher than that expected for the clinical formulation image required for efficacy in humans. For orally administered drugs, the dosing formulation is typically administered by oral gavage in order to ensure that a fixed volume of the dose is consistently delivered over the duration of the study. Consequently, high concentration solution or suspension formulations are highly desired. At these high doses, limitations to oral absorption may become apparent due to either solubility or dissolution rate-limited processes (Fahr and Liu, 2007; Stegemann *et al.*, 2007). These issues become exacerbated with compounds that exhibit poor aqueous solubility. To meet these challenges, enabling or solubilizing formulation technologies is often required to deliver sufficient quantities of drug. However, in order to fully resolve potential adverse effects related solely to the drug candidate and avoid complications due to effects from other components present in the formulation, it is important to limit the type and concentration of excipients used in the formulations to avoid ambiguous results. Consequently, in this setting it is preferred to utilize formulation approaches that require minimal excipient loading and biologically inert excipients. This can limit the use of solubilizing agents such as organic cosolvents, lipid emulsions, or complexing agents such as cyclodextrins due to potential toxicity observed related to the large amount of excipients typically required for these approaches.

Despite these challenges, there are several enabling formulation technologies that have emerged in downstream development that avoids these issues, but it can be a challenge to implement these approaches in the discovery space due to the limitations in drug availability as well as rapid turnaround times. In this chapter, case studies will be presented demonstrating the application of specific enabling formulation technologies to support the delivery of discovery drug candidates for early safety toxicology studies. For compounds with poor aqueous solubility exhibiting limited absorption, formulations such as nano-suspensions and amorphous solid dispersions have emerged as highly successful approaches for achieving sufficient exposure margins without requiring high concentrations of undesired excipients. Importantly, these approaches also have a line of sight toward formulation strategies used later in the clinic, reducing the time and resources required for formulation development.

6.2 General

For the formulation scientist, it is highly desirable to obtain an understanding of the properties of a potential new drug candidate during drug discovery lead optimization in order to help guide the appropriate formulation strategy at each stage. This involves characterizing the physicochemical properties of the drug molecule such as identifying a suitable solid form, solubility, and dissolution rate, as well as molecular properties such as molecular weight, pK_a, $\log P/D$, polar surface area, and permeability. It is critical for formulation scientists to engage with medicinal chemists to obtain sufficient material to allow for early characterization of new chemical matter that may become new drug candidates.

At the same time, it is also important to obtain an understanding of the biopharmaceutical performance of the new drug candidate to identify potential risks for poor absorption due to factors such as poor solubility and permeability, particularly at high doses. Low dose pharmacokinetic (PK) studies can give an assessment of the overall absorption, distribution, metabolism, and excretion (ADME) profile of the compound after being dosed *in vivo*. Biopharmaceutical challenges for absorption are expected to increase as the dose is raised to support high-dose safety toxicology studies. For early GLP toxicology studies, ICH topic M3(R2) guidelines for nonclinical safety studies to support human clinical trials require large exposure multiples obtained using either the MTD or maximum feasible dose (MFD) (ICH, 2009). Thus, it is important to develop a suitable formulation strategy to administer and deliver a high dose of drug. This can be a significant challenge for compounds with poor solubility and appropriate enabling formulation approaches are required to mitigate those risks.

6.3 Nanosuspension Formulations

One enabling formulation approach for poorly soluble drugs is the reduction in particle size of the solid material to the nanometer scale (Kesisoglou *et al.*, 2007; Merisko-Liversidge and Liversidge, 2008; Merisko-Liversidge *et al.*, 2003; Pu *et al.*, 2009; Rabinow, 2004). Generally, as the size of the drug particles is reduced, the original crystalline phase of the material is maintained, resulting in the formation of nanocrystals of the starting material. These products typically appear as milky colloidal suspensions of drug nanoparticles in aqueous vehicles and are known as nanosuspension formulations. These formulations can be handled similarly to solutions and are well suited for oral gavage administration to support early safety toxicology studies.

The primary advantage of nanosuspension formulations is that the small size of the nanoparticles results in a higher exposed surface area for the drug solid that proportionately increases the dissolution rate of the compound, as expressed by the Noyes–Whitney equation (Equation 6.1) (Noyes and Whitney, 1897)

$$\frac{dC}{dt} = \frac{AD(C_s - C)}{h} \tag{6.1}$$

where A is the exposed surface area of the particle, D is the diffusion coefficient of the compound, h is the thickness of the boundary layer between the dissolving solid and the solvent, and $(C_s - C)$ is the concentration gradient of the compound between the particle surface and the bulk solvent. As can be seen by this relationship and since total surface area is inversely proportional to the radius of the particle, smaller drug particles have faster dissolution rate.

In addition, a slight increase in the solubility of the drug solid can occasionally be observed as the particle size is reduced, as described by the Freundlich–Ostwald equation (Equation 6.2) (von Helmholtz, 1886)

$$S = S_\infty \exp\left(\frac{2\gamma M}{r \rho RT}\right) \tag{6.2}$$

where S_∞ is the solubility of an infinitely large particle, γ is the interfacial surface tension, M is the molecular weight of the compound, r is the radius of the particle, ρ is the density of the particle, R is the gas constant, and T is the temperature. Due to this relationship, a moderate increase of solubility around 10–15% may be expected to be seen for a typical drug molecule of 500 g/mol molecular weight, ρ of 1 g/mL, and γ value of 15–20 mN/m when the particle size of the drug is reduced to 100 nm compared to a typical unmilled, micron-sized solid material (Kesisoglou and Mitra, 2012). Furthermore, in some cases the creation of high energy exposed surfaces as well as defects to the crystalline lattice during the particle size reduction process can also result in increased solubility, which has been observed in certain cases (Muller and Peters, 1998). Both of these effects can result in an enhancement in absorption and bioavailability compared with a conventional suspension of the starting material consisting of large solid particles (Kesisoglou et al., 2007; Merisko-Liversidge and Liversidge, 2011).

Since the drug particles in these nanosuspension formulations typically remain crystalline, chemical stability is usually maintained, particularly when compared with higher energy solid forms such as amorphous formulations. This can help enable support for studies with long duration without risk of chemical degradation of the drug molecule. However, physical stability of the nanoparticles remains a concern, and prevention of colloidal instability risks such as aggregation, agglomeration, or Ostwald ripening is critical. In order to address these risks, polymer and/or surfactant excipients are added during processing that adsorb to the surface of the nanoparticles to stabilize the material as discrete particles. Fortunately, typically only a small amount of these stabilizing excipients are required for maintaining physically stable nanosuspension formulations, making them particularly advantageous for toxicological evaluation (Kesisoglou and Mitra, 2012; Muller et al., 2011). These excipients include biologically inert polymers such as naturally derived cellulosics (e.g., hydroxypropyl cellulose and hypromellose), povidones (PVP), and poloxamers, as well as surfactants such as polysorbates and sodium dodecyl sulfate (SDS), among others (van Eerdenbrugh et al., 2009a). These additives adsorb to the surface of the solid drug particles to improve the milling process and to stabilize the discrete nanoparticles by providing a steric or ionic barrier to inhibit aggregation or agglomeration (Merisko-Liversidge and Liversidge, 2011; Peltonen and Hirvonen, 2010; van Eerdenbrugh et al., 2009a, b; Wu et al., 2011). However, it is often difficult to determine the optimal identity and concentration of the additives required to effectively stabilize the formulations prior to formulation development. The addition of nonoptimal stabilizers may result in inefficient milling, aggregation or agglomeration of the nanoparticles, or Ostwald ripening. As a result, formulation optimization of a nanosuspension is typically empirical and relies upon the evaluation of multiple formulation compositions to identify a suitable option (van Eerdenbrugh et al., 2009b).

Particle size reduction is typically performed using a top-down milling approach, and this approach has been undertaken on the commercial scale for a number of drug products such as Emend[*] and Rapamune[*], among others

(Merisko-Liversidge and Liversidge, 2011; Oliver *et al.*, 2007; Shen and Wu, 2007; Wu *et al.*, 2004). These approaches utilize high energy, high shear processes such as wet milling or homogenization (Merisko-Liversidge and Liversidge, 2011). However, these techniques typically require a large amount of drug material and are difficult to scale down. As a result, it can be time and material consuming to develop a suitable nanosuspension formulation, particularly within the constraints of the discovery and early development space.

In order to facilitate the use of nanosuspension formulations in discovery, resonant acoustic mixing has recently been discovered as an alternative and highly effective way of preparing nanosuspension formulations on a very small scale (as low as single milligram) (Leung *et al.*, 2014a, b). In addition, the acoustic mixing approach has enabled the development of a rapid, material-sparing parallel screen for optimal formulation parameters using a standard 96-well microtiter plate suitable for application in the discovery space (Figure 6.1). The particle size of formulations with varying compositions can be tracked over time in a 384-well plate using a high-throughput dynamic light scattering instrument (Figure 6.2). Since acoustic mixing is a highly homogeneous process, the selected nanosuspension formulations can be scaled up with little loss in efficiency. Accordingly, optimal formulation parameters can be selected from the initial screening approach and scaled up to support *in vivo* studies. As a result, formulation optimization can be performed rapidly and a suitable nanosuspension formulation can be prepared for use in preclinical studies.

6.3.1 Case Study 1: Nanosuspension Formulation for Increased Maximum Dose and Systemic Drug Exposure

As an example of this formulation strategy, compound A was identified as a candidate for progression to development. Characterization of compound A indicated high permeability in LLC-PK1 cells ($P_{app} = 37 \times 10^{-6}$ cm/s) but poor solubility (<50 µg/mL) in the physiological pH range for its crystalline freebase despite low lipophilicity as measured by log$D$ at pH 7 (Table 6.1). The solubility of compound A was low in simulated biorelevant fluids including simulated gastric fluid (SGF, pH 2) as well as fasted state simulated intestinal fluid (FaSSIF, pH 6.5) (Dressman *et al.*, 1998; Jantratid and Dressman, 2009). As a consequence, the compound carried a risk of solubility-limited absorption at high doses and a likely need for enabling formulations in preclinical safety studies.

Serendipitously, early batches of compound A prepared in discovery consisted of crystalline particles in the nanometer range (100–200 nm) without particular efforts to control particle size upon crystallization. Suspensions of such nanoscale API in 10% polysorbate 80 were successfully used for early *in vivo* experiments in rodents since they provided adequate exposures in these studies. However, since solubility-limited absorption was anticipated to be a risk as the compound progressed to studies involving higher doses, enabling

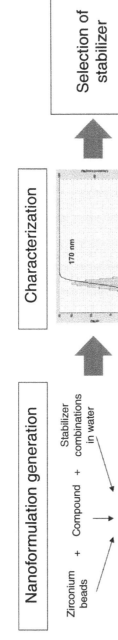

Figure 6.1 Selection of a stabilizer for crystalline nanosuspensions of compound A.

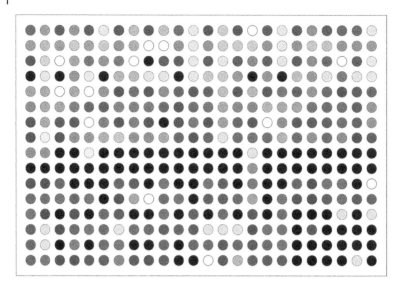

Figure 6.2 Visual heat map of particle sizes following acoustic mixing of a sample of naproxen with varying polymer and surfactant excipients over time, where green indicates small nanoparticles, red indicates larger nanoparticles, and black indicates lack of stable nanoparticles. (*See insert for color representation of the figure.*)

Table 6.1 Physicochemical properties of compound A.

Property	Value
Physical form	Crystalline freebase
Molecular weight (g/mol)	550.1
Log$D$ at pH 7	2.47
pK_a	4.78
Buffer solubility (µg/mL):	
pH 2 (0.01 N HCl)	0.10 (measured pH 1.8)
pH 4 (50 mM citrate buffer)	0.10 (measured pH 3.9)
pH 6 (50 mM citrate buffer)	0.39 (measured pH 6.1)
pH 8 (50 mM phosphate buffer)	48 (measured pH 7.9)
pH 10 (50 mM carbonate buffer)	>2000 (measured pH 9.9)
SGF solubility (µg/mL)	0.02 (measured pH 1.8)
FaSSIF solubility (µg/mL)	10 (measured pH 6.5)
LLC-PK1 Papp ($\times 10^{-6}$ cm/s)	37

and solubilizing formulations such as amorphous solid dispersions and aqueous solutions of the *in situ* sodium salt of compound A were investigated. These formulations gave comparable exposures to that for the nanosuspension of the crystalline free form. However, a high dose level of 1000 mg/kg was required in order to achieve the required exposure margins. Both the amorphous solid dispersion and *in situ* sodium salt formulations were not considered viable for administration at this high dose level due to limitations in their maximum feasible concentration (high viscosity and unacceptable pH, respectively). In addition significant PK variability was observed for the two formulations, particularly for the solution of the sodium salt (likely caused by uncontrolled precipitation in acidic gastric environment). As a result, nanosuspensions of submicron crystalline material, as used in early *in vivo* studies, were considered the preferred option for GLP toxicology studies due to the high doses required.

Input from development colleagues was collected at that time, and challenges in reproducing submicron material on scale-up as well as processability issues were highlighted as significant developability risks. A batch of compound A considered to be more representative of larger-scale GMP material (micron-size particles) was generated and provided to the discovery team for *in vivo* evaluation in a mouse model. A particle size effect study was conducted in order to compare its bioperformance with that of the original nanoscale material. The experiment, conducted at doses of 300, 750, and 1500 mg/kg, highlighted a major impact of particle size on absorption with a greater than threefold difference in area under the curve (AUC) exposures for the concentrations of API in plasma between the two particle size distributions at all doses (Figure 6.3). Importantly, the nanosuspension formulation could be successfully prepared and administered at a top dose of 1500 mg/kg while maintaining high exposure levels. Based on this result, a nanosuspension formulation generated from micron-size GLP material was desired for rodent GLP safety studies in order to enable high doses and optimize safety exposure multiples.

Nanosuspension formulation screening was conducted as described in the previous section using acoustic mixing on a 96-well plate format to identify a stabilizer system that provided optimal milling properties (particle size, milling time) as well as suitable stability. Milling of the material in the presence of 10% polysorbate 80 gave a particle size distribution similar to that for other excipients tested as well as an excellent physical stability. No significant changes in particle size of the nanosuspension were observed over time during storage. In addition, freeze–thaw cycling did not induce changes in particle size. Because early *in vivo* experiments were conducted on suspensions of nanoscale material in 10% polysorbate 80, bridging PK studies were not needed and 10% polysorbate 80 was selected as the nanosuspension formulation composition for GLP studies. Scale-up of the nanosuspension formulation was subsequently conducted using the acoustic mixing process to generate sufficient material for GLP safety studies.

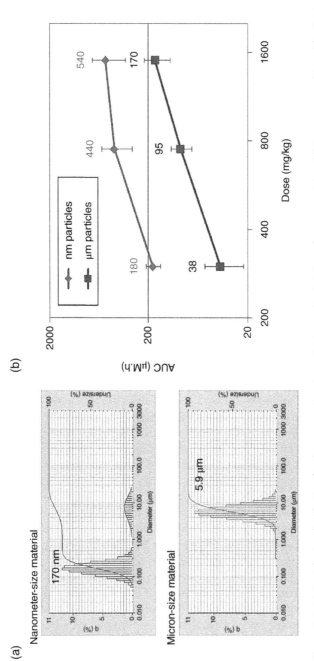

Figure 6.3 (a) Particle size distribution for the nanometer-size and micron-size batches of compound A and (b) exposures achieved for the two batches in the particle size effect study when dosed *in vivo* in a mouse model.

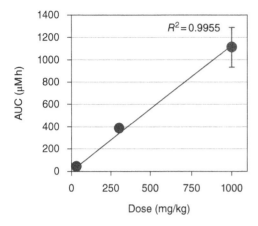

Figure 6.4 AUC versus dose plot for compound A administered in GLP toxicology studies in a mouse model.

GLP toxicology studies were conducted in a mouse model, and high exposures were achieved using the nanosuspension approach, with mean sex-combined AUC_{0-24h} highlighting excellent dose proportionality from 30 to 1000 mg/kg/day (Figure 6.4). The exposures, consistent with that achieved in early discovery, enabled suitable exposure multiples to progress in the safety study.

In this example, while an amorphous solid dispersion or *in situ* salt formulation was also considered, a nanosuspension strategy provided clear advantages since it could be successfully dosed at 1000 mg/kg, while providing dose proportional increases in exposure upon oral administration. This case study also highlights how a nanosuspension formulation can help provide a direct line of sight from early discovery studies into later stage safety toxicology studies. Successful nanomilling of later batches of compound A provided material that behaved substantially similar to early batches of compound A prepared in discovery. As a result, it can be seen that it is critical for the formulation scientist to be fully engaged with teams throughout the drug discovery process.

6.4 Amorphous Solid Dispersion Formulations

During development of a new drug candidate, the drug substance is preferably isolated in the solid state as an ordered, crystalline material. This generally offers advantages for ease in isolation and purification as well as improved chemical and physical stability in the solid state for long-term storage. However, these advantages can come with a concomitant limitation in solubility as additional energy is required to break the lattice energy of the crystalline form to fully dissolve the solid material. This can restrict the dissolution rate as well as the overall solubility of a drug solid, potentially resulting in poor absorption at the high doses required for safety toxicology studies.

In contrast to the well-ordered structure of a crystalline lattice, the amorphous form of a drug substance consists of a disordered, random structure. This results in a higher ground state free energy for the solid material. Since there is no longer an energy cost required to break a crystalline lattice, amorphous materials typically have faster dissolution rates and higher solubilities compared with the corresponding crystalline form (Alonzo *et al.*, 2010; Craig, 2002; Leuner and Dressman, 2000; van den Mooter, 2012). This effect has been termed a "spring" to help boost the initial dissolution and solubility of a drug solid, which can help improve absorption and the overall biopharmaceutical performance of the material (Guzman *et al.*, 2007). However, despite these advantages, due to the inherent high free energy state of amorphous material, this is a higher risk of chemical instability as well as physical stability risks such as conversion to a more thermodynamically stable crystalline material with a correspondingly lower solubility and absorption profile.

In order to take advantage of this increase in solubility while mitigating these stability risks, amorphous solid dispersions have emerged as another enabling formulation technology and involve the stabilization of the amorphous form of a drug solid within a polymer matrix (Alonzo *et al.*, 2010; Baghel *et al.*, 2016; Craig, 2002; He and Ho, 2015; Leuner and Dressman, 2000; van den Mooter, 2012; Vasconcelos *et al.*, 2007). This effect can be the result of several factors. Physical entrainment within the polymer network can result in decreased mobility of the drug molecules and reduced tendency for nucleation and crystallization to occur. Below the glass transition temperature (T_g), the material effectively exists as a frozen glass with restricted mobility. This effect can be measured using modulated differential scanning calorimetry (mDSC) to detect thermal transitions associated with effects such as the T_g, as well as by solid-state nuclear magnetic resonance (ssNMR) spectroscopy to directly measure the molecular mobility of the entrained drug molecule (Paudel *et al.*, 2014). Related to this, the "solubility" of the drug molecule within the polymer can be another factor involved in stabilization of the amorphous form as a kind of solid solution (Higashi *et al.*, 2015). There may be specific and nonspecific interactions such as H-bonding, electrostatic and hydrophobic attraction, and van der Waals forces between the drug and polymer helping to stabilize the amorphous form of the drug.

Another risk with formulations of amorphous material is crystallization of the drug after dissolution either in the dosing vehicle or *in vivo*. Since the amorphous form of a drug has higher solubility compared with a crystalline analog, the drug molecule exists in a supersaturated state with respect to the crystalline form and may crystallize as a more thermodynamically stable, lower solubility material. Despite this risk, the polymers used in amorphous solid dispersions not only stabilize the amorphous form of the drug in the solid state but can also often facilitate prolonged supersaturation of the higher solubility amorphous form after dissolution in aqueous media, a phenomenon coined as the "parachute" effect (Brouwers *et al.*, 2009; Guzman *et al.*, 2007; Raina *et al.*, 2014; Warren *et al.*, 2010; Xu and Dai, 2013). Depending upon the polymer selected, nucleation, crystallization, and precipitation of the supersaturated drug may be inhibited for a sufficient amount of time to allow for the formulation to be dosed and absorbed *in vivo*. This combined "spring and parachute" effect makes amorphous solid dispersions an attractive approach to increasing the solubility and absorption of poorly soluble compounds.

Although the drug loading within the polymer matrix is typically limited to 20–40% due to these stability risks, the polymers used are typically non-absorbed and biologically inert, avoiding issues with adverse effects related to the excipients used in the formulation composition (Kadajji and Betageri, 2011; Teja *et al.*, 2013). Naturally derived cellulosic polymers such as hypromellose acetate succinate (HPMCAS) have proven to be effective polymer excipients for amorphous solid dispersions (Curatolo *et al.*, 2009; Friesen *et al.*, 2008). As a result, due to the combination of improved solubility for the drug and use of relatively safe polymer excipients, amorphous solid dispersions are ideal formulations for safety toxicology studies.

6.4.1 Case Study 2: Amorphous Solid Dispersion Formulation for Resolving Complex Polymorphism Challenges

In the second case study, compound B was initially isolated as an amorphous freebase that was used in early discovery studies without having any *in vivo* exposure issues. However, when the compound progressed toward consideration for development as a clinical candidate, phase screening was initiated in order to identify a stable crystalline form for development. It was quickly found that compound B exhibited extremely complicated polymorphism phenomena with limited time to identify the most thermodynamically stable form. It was also found that the solubility of the various crystalline forms was decreased by 30-fold compared with the initial amorphous phase, resulting in a serious risk of conversion of the amorphous form to a less soluble crystalline form and impacting oral absorption. In order to move the program forward without delay due to the time required for resolving phase definition and to derisk any potential decrease in exposure in preclinical safety studies, several enabling formulation strategies were pursued, including an amorphous solid dispersion approach.

In this case, a material-sparing high-throughput polymer screen involving solvent casting in a 96-well plate was used for selecting and optimizing the formulation components (Barillaro *et al.*, 2008; Chiang *et al.*, 2012; Manksy *et al.*, 2007; Moser *et al.*, 2008; Shanbhag *et al.*, 2008). A total of 10 mg of API was used with a variety of different polymer and surfactant excipients. The API and excipients were initially solubilized in organic solvent such as acetone or methanol, dispensed into separate wells in the 96-well plate, and then rapidly evaporated to leave a dried film. The biorelevant fluid FaSSIF was then added in, and the solubility of compound B was analyzed from each film over time to give an estimate for the solubility and dissolution profile of the compound at the site of absorption *in vivo* (Dressman *et al.*, 1998; Jantratid and Dressman, 2009). The solubility and dissolution summary is presented in Figure 6.5. Among all the different drug loads and excipient compositions tested, many achieved solubility in FaSSIF >600 μg/mL compared with the film consisting of amorphous compound B alone at below 400 μg/mL.

In order to ensure prolonged duration of maintaining oversaturated solubility in the gastrointestinal (GI) tract, additional anti-nucleation agents were also screened by using a 96-well plate high throughput screening method (Vandercruysa *et al.*, 2007). The two selected amorphous solid dispersion formulations were transferred into the wells and vehicles consisting of different excipients in various percentages, and combinations were added subsequently. Biorelevant FaSSIF media was then added to each well and the samples were stirred at 37°C. After an appropriate amount of time, the plate was centrifuged, and the residual solid material was examined for the presence of any conversion

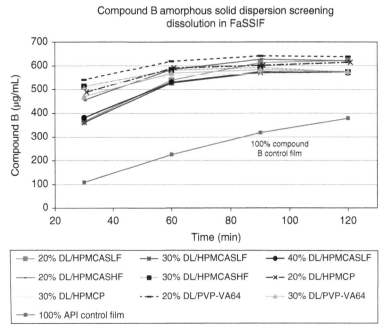

Figure 6.5 Solubility in FaSSIF of high-throughput amorphous solid dispersion formulation screening of a variety of polymers and drug loadings.

Table 6.2 Oral exposure of compound B in rat after 150 mg/kg dose.

PK 100 mg/kg in rat	C_{max} (μM)	T_{max} (h)	$AUC_{(0-x)}$ (μM h)	Exposure multiple	Physical stability (in vehicle)
Partially solubilized formulation of the HCl salt with surfactants	48	2	611	127×	Unstable (disproportionation of hydrochloride salt)
Nanosuspension of crystalline freebase (wet milling)	21.5	2	254	53×	Unstable (converting to a new phase)
Amorphous solid dispersion of freebase (spray drying)	43.8	2.3	686	143×	Stable (stay as amorphous freebase)

to crystalline material exhibiting birefringence under cross-polarized optical microscopy. By using this method, 96 different vehicle combinations were screened for the two amorphous solid dispersion formulations. The results indicated that the HPMCAS-HF amorphous solid dispersion was more resistant to crystallization compared with HPMCAS-LF and provided more flexibility in the selection of anti-nucleating agent in the suspending vehicles. Based on this screening, amorphous solid dispersion of 20% drug load in HPMCAS-LF and 20% drug load in HPMCAS-HF were selected for scale-up. For the HPMCAS-HF amorphous solid dispersion, Poloxamer 188, Kollidon® VA64, PVP, and HPMC were identified as optimal anti-nucleation excipients to prevent crystallization in the dosing vehicle (Vandercruysa *et al.*, 2007).

The performance of various formulations of compound B *in vivo* confirmed that sufficient oral exposure was achieved by three different enabling or solubilizing formulation approaches as shown in Table 6.2. However, only the amorphous solid dispersion formulation showed sufficient physical stability in the dosing vehicle. Due to the complex polymorphism and phase map of the compound, it was challenging to maintain a stable and consistent physical form in the formulation. The formulations using the surfactant solution of the HCl salt and the nanosuspension of the crystalline freebase showed phase conversion in the dosing vehicle. In contrast, the amorphous solid dispersion prevented conversion of the API to other physical forms and maintained the drug in a stable amorphous form. In this case, not only did the amorphous solid dispersion approach enable high exposure margins, but it also provided enhanced physical stability for the drug substance. As a result, it was selected to support the early preclinical toxicology formulation to move the program forward without requiring a delay in order to resolve the complex polymorphism issue.

6.4.2 Case Study 3: Amorphous Solid Dispersion Formulation for Improving Oral Absorption

In the third case study, compound C was identified as a promising new drug candidate. Compound C was an insoluble compound with poor solubility for the crystalline freebase in aqueous buffers across the physiological pH range (<0.5 μg/mL solubility in a pH range from 1.5 to 8.0), as well as in simulated biorelevant media with a solubility <0.5 μg/mL in SGF, 1.2 μg/mL in FaSSIF, and 7.5 μg/mL in fed state simulated intestinal fluid (FeSSIF). It also had modest permeability in LLC-PK1 cells ($P_{app} = 7.9 \times 10^{-6}$ cm/s) and low clearance in rats, dogs, and monkeys. The combination of poor solubility and modest permeability suggested that the compound would likely encounter significant absorption risks upon oral administration, and a tentative BCS classification of IV was assigned. The physicochemical properties of compound C are summarized in Table 6.3.

In order to evaluate the factors potentially limiting oral absorption, the microscopic mass balance approach (MiMBA) (Oh *et al.*, 1993; Rohrs, 2006) was applied in the early discovery stage on compound C. The MiMBA approach is a biopharmaceutical model developed by Amidon and coworkers to estimate the fraction absorbed (F_a) of a given dose. The solubility (represented by dose number or D_o), dissolution rate (represented by dissolution number or D_n), and permeability (represented by absorption number or A_n) of a given compound are used to estimate F_a using scaling factors for the appropriate species. The interplay of these properties gives rise to contour maps that portray the

Table 6.3 Physicochemical properties of compound C.

Property	Value
Physical form	Crystalline freebase
Molecular weight (g/mol)	614
Log$D$ at pH 7	3.8
pK_a	<2 (basic groups)
Buffer solubility (μg/mL): pH 1.5–8.0	<LOQ[a]
SGF solubility (μg/mL)	<LOQ[a]
FaSSIF solubility (μg/mL)	1.2 (measured at pH 6.2)
FeSSIF solubility (μg/mL)	7.5 (measured at pH 4.7)
LLC-PK1 Papp ($\times 10^{-6}$ cm/s)	7.9
BCS class	IV

[a] LOQ = 0.5 μg/mL.

"biopharmaceutics landscape" of a compound. Since these properties can be measured relatively early in discovery, MiMBA modeling can help predict particular liabilities that may influence the bioperformance for a compound early on and can be critical for determining an appropriate formulation strategy.

The MiMBA model was applied to compound C and the F_a contour plots are presented in Figure 6.6. As illustrated in the MiMBA simulation for the bioperformance of compound C in rats and dogs (Figure 6.6a and b) at a tentatively selected dose of 10 mg/kg, the oral absorption of compound C is predicted to be limited with <10% of the dose expected to be absorbed with further loss in performance as the dose is increased. This limitation was suggested to be due to the low solubility of compound C. Increased F_a was predicted to be seen with lower D_o (increased solubility), while reduction in particle size to raise D_n (increased dissolution rate) did not appear to significantly improve F_a. In this case, enabled formulations that improved the solubility (such as amorphous solid dispersions) rather than formulations that improved dissolution rate (such as nanosuspensions) of compound C were likely needed for preclinical studies.

As a result, a material-sparing high-throughput polymer screen for an amorphous solid dispersion of the API was conducted as described in case study 2. Based on the results obtained, hydroxypropyl methylcellulose phthalate (HPMCP) was identified as the most suitable polymer for the amorphous solid dispersion. In order to scale up and generate sufficient quantities of the amorphous solid dispersion to support *in vivo* studies, spray drying was selected as the preparation technique. Spray drying is an efficient method for producing amorphous solid dispersions. The drug and a selected polymer are co-dissolved in a volatile organic solvent before being sprayed through a nozzle. The solution is atomized into fine droplets where the solvent is rapidly evaporated under a stream of hot gas before the drug has had sufficient time to phase separate from the polymer and crystallize. The resulting amorphous solid dispersion material can be collected and used.

In this case, a mixture of compound C and HPMCP at drug loadings of either 30 or 40% were fully dissolved in a solution consisting of 9:1 acetone:water and was spray dried using a ProCepT micro-spray dryer (Ormes *et al.*, 2013). The properties of the resulting amorphous solid dispersions are summarized in Table 6.4. The materials produced after spray drying were found to be amorphous with a T_g of close to 140°C for the amorphous solid dispersions consisting of 30 and 40% drug loading. In addition, both amorphous solid dispersion materials had significantly higher solubility in FaSSIF that endured over a 4 h time period compared with the original crystalline material (1.2 μg/mL) as well as good physical stability. The enhancement in solubility was expected to improve the F_a and bioperformance of compound C.

The amorphous solid dispersion containing 40% drug load in HPMCP was selected for evaluation *in vivo* in rats in a dosing vehicle consisting of 0.5% methylcellulose containing 5 mM hydrochloric acid and 0.25% SDS. The oral

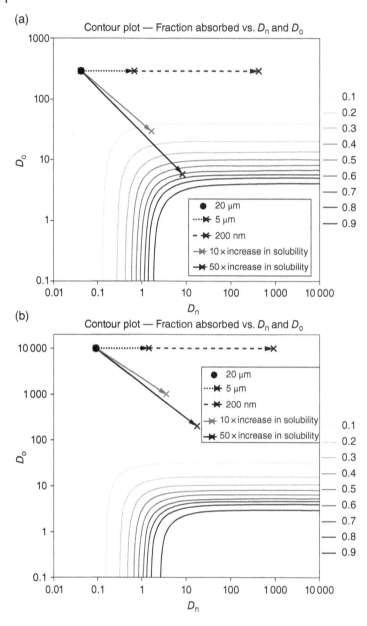

Figure 6.6 (a) MiMBA prediction in rats at 10 mg/kg and (b) MiMBA prediction in dogs at 10 mg/kg. The particle size was set to be 20 μm initially, 5 μm post jet-milling, and 200 nm post nanomilling for the simulation. The two scenarios of solubility improvement at 10- and 50-fold were simulated.

Table 6.4 Properties of spray dried amorphous solid dispersion of compound C.

Property	30% drug load	40% drug load
Crystallinity	Amorphous	Amorphous
Glass transition temperature (°C)	139	140
FaSSIF solubility after 4h (µg/mL)	81	69

Table 6.5 Oral exposure of compound C in rats after 100 mg/kg dose.

Formulation	Dose (mg/kg)	$AUC_{0-\infty}$ (µM h)	C_{max} (µM)
Suspension of crystalline freebase in 10% polysorbate 80	100	32.9	3.3
Amorphous solid dispersion	10	34.3	2.7
Amorphous solid dispersion	100	101	6.7
Amorphous solid dispersion	300	198	9.2

exposure of compound C was improved approximately threefold in AUC compared with the original crystalline API formulated as a suspension in 10% polysorbate 80 (Table 6.5). However, as the dose was increased, less than proportional responses were observed and the overall exposures were lower than expected. This may be due to solubility limitations at the high doses despite the use of the amorphous solid dispersion formulation. Nevertheless, sufficient exposures were achieved and the formulation was selected for safety toxicology studies in rats.

While sufficient exposures were observed in rats, the MiMBA model suggested a high risk in bioperformance in dogs even with a 50-fold increase in solubility at a low dose of 10 mg/kg (Figure 6.6b). The bioperformance was expected to drop further as the dose is increased. The oral exposure is expected to be more limited by solubility at higher doses, and additional solubilization is likely required to achieve sufficient oral exposure in safety studies. The following two approaches were proposed to further enhance the solubility in the GI tract:

1) Addition of surfactants in the dosing vehicle to improve the drug solubility of the amorphous solid dispersion
2) Lowering of the drug load in the amorphous solid dispersion to maintain longer supersaturation in the GI tract

Accordingly, in order to guide formulation selection for drug safety studies in dogs, three further amorphous solid dispersion formulations evaluating these

Table 6.6 Compositions of solubilized amorphous solid dispersion formulations of compound C.

Formulation	Dosing vehicle
Amorphous solid dispersion (30% drug load/70% HPMCP)	10% Polysorbate 80 in 0.5% methylcellulose acidified to pH 4
Amorphous solid dispersion (40% drug load/60% HPMCP)	10% Polysorbate 80 in 0.5% methylcellulose acidified to pH 4
Amorphous solid dispersion (40% drug load/60% HPMCP)	0.5% Methylcellulose containing 5 mM hydrochloric acid and 0.25% sodium lauryl sulfate

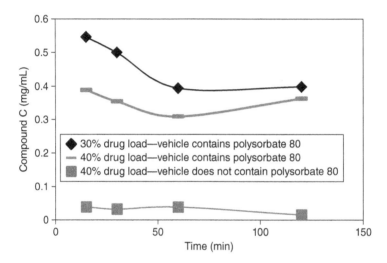

Figure 6.7 Dissolution profile of compound C from three solubilizing-enabled formulations.

approaches were prepared (Table 6.6). The formulations were prepared at a concentration of 20 mg/mL of compound C for dissolution testing. The formulations were first diluted twofold with SGF, and the resulting mixtures were then diluted fourfold with FaSSIF after 15 min in order to simulate the effects of oral administration. The dissolution profiles in FaSSIF were plotted in Figure 6.7. As indicated by the dissolution profiles, the concurrent addition of polysorbate 80 in the dosing vehicle provided a dramatic increase in the solubility of compound C. Furthermore, lowering of the drug load in the amorphous solid dispersion from 40 to 30% also provided an enhancement in solubility.

Based on these results, the amorphous solid dispersion with 30% drug load suspended in the dosing vehicle containing polysorbate 80 was selected for

Table 6.7 Oral exposure for the selected formulation of compound C in fed dogs: The formulation consisted of an amorphous solid dispersion consisting of 30% drug load of compound C in HPMCP in a dosing vehicle containing 10% polysorbate 80 in 0.5% methylcellulose at pH 4.

Dose (mg/kg)	$AUC_{0-\infty}$ ($\mu M\,h$)	C_{max} (μM)
30	204	8.3
100	356	12
300	734	17

evaluation *in vivo* in dogs. In this case, a close to dose proportional response was achieved as the dose was increased from 30 to 300 mg/kg with this formulation approach (Table 6.7). Sufficient exposures were achieved and this formulation approach was chosen for safety toxicology studies in dogs.

In this example, an early understanding of the physicochemical properties of compound C enabled a prediction of the potential liabilities that may affect bioperformance. Solubility-limited absorption was identified as a significant risk, and an amorphous solid dispersion formulation approach was used to support the compound for high dose *in vivo* studies. This approach was suitable for the rat model, but further solubilization was required for the dog model. The addition of surfactants to the dosing vehicle of the amorphous solid dispersion formulation allowed sufficient exposures to be achieved at high dose.

6.5 Conclusion

Obtaining an early understanding of the safety and toxicity profile of a new drug candidate can help reduce unanticipated risks in the clinic and may help to reduce the high attrition rate of drugs in clinical development. As part of this assessment, *in vivo* safety evaluation of new drug candidates is critical and requires the use of suitable formulation approaches. These studies require consistent and reproducible formulations that support a wide dose range using low concentrations of safe or inert excipients. Due to the high dose levels required, enabling formulation technologies may be required to ensure that sufficient absorption and exposure margins are obtained for compounds with challenging physicochemical properties. For formulation scientists, it can be challenging to implement enabling formulation approaches in the discovery space due to time and material limitations, but as can be seen by the case studies presented here, recent advances in nanosuspension and amorphous solid dispersion strategies have demonstrated that success can be achieved.

References

Alonzo, D. E.; Zhang, G. G. Z.; Zhou, D.; Gao, Y.; Taylor, L. S., Understanding the behavior of amorphous pharmaceutical systems during dissolution. *Pharm. Res.* 2010, 4, 608–618.

Baghel, S.; Cathcart, H.; O'Reilly, N. J., Polymeric amorphous solid dispersions: a review of amorphization, crystallization, stabilization, solid-state characterization, and aqueous solubilization of biopharmaceutical classification II drugs. *J. Pharm. Sci.* 2016, 105, 2527–2544.

Barillaro, V.; Pescarmona, P. P.; Van Speybroeck, M.; Thi, T. D.; Van Humbeeck, J.; Vermant, J.; Augustijns, P.; Martens, J. A.; Van Den Mooter, G., High-throughput study of phenytoin solid dispersions: formulation using an automated solvent casting method, dissolution testing, and scaling-up. *J. Comb. Chem.* 2008, 10, 637–643.

Bass, A. S.; Cartwright, M. E.; Mahon, C.; Morrison, R.; Snyder, R.; McNamara, P.; Bradley, P.; Zhou, Y.-Y.; Hunter, J., Exploratory drug safety: a discovery strategy to reduce attrition in development. *J. Pharmacol. Toxicol. Methods* 2009, 60, 69–78.

Blomme, E. A. G.; Will, Y., Toxicology strategies for drug discovery: present and future. *Chem. Res. Toxicol.* 2016, 29(4), 473–504.

Bowes, J.; Brown, A. J.; Hamon, J.; Jarolimek, W.; Sridhar, A.; Waldron, G.; Whitebread, S., Reducing safety-related drug attrition: the use of *in vitro* pharmacological profiling. *Nat. Rev. Drug Discov.* 2012, 11, 909–922.

Brouwers, J.; Brewster, M. E.; Augustijns, P., Supersaturating drug delivery systems: the answer to solubility-limited oral bioavailability? *J. Pharm. Sci.* 2009, 98, 2549–2572.

Chiang, P.-C.; Ran, Y.; Chou, K.-J.; Cui, Y.; Sambrone, A.; Chan, C.; Hart, R., Evaluation of drug load and polymer by using a 96-well plate vacuum dry system for amorphous solid dispersion drug delivery. *AAPS PharmSciTech* 2012, 13, 713–722.

Craig, D., The mechanisms of drug release from solid dispersions in water-soluble polymers. *Int. J. Pharm.* 2002, 231, 131–144.

Curatolo, W.; Nightingale, J. A.; Herbig, S. M., Utility of hydroxypropylmethylcellulose acetate succinate (HPMCAS) for initiation and maintenance of drug supersaturation in the GI milieu. *Pharm. Res.* 2009, 26, 1419–1431.

Dressman, J. B.; Amidon, G. L.; Reppas, C.; Shah, V. P., Dissolution testing as a prognostic tool for oral drug absorption: immediate release oral dosage forms. *J. Pharm. Sci.* 1998, 1, 11–22.

van Eerdenbrugh, B.; Stuyven, B.; Froyen, L.; van Humbeeck, J.; Martens, J. A.; Augustijns, P.; van den Mooter, G., Downscaling drug nanosuspension production: processing aspects and physicochemical characterization. *AAPS PharmSciTech* 2009a, 10, 44–53.

van Eerdenbrugh, B.; Vermant, J.; Martens, J. A.; Froyen, L.; van Humbeeck, J.; Augustijns, P.; van den Mooter, G., A screening study of surface stabilization during the production of drug nanocrystals. *J. Pharm. Sci.* 2009b, 98, 2091–2103.

Fahr, A.; Liu, X., Drug delivery strategies for poorly water-soluble drugs. *Expert Opin. Drug Deliv.* 2007, 4, 403–416.

Friesen, D. T.; Shanker, R.; Crew, M.; Smithey, D. T.; Curatolo, W. J.; Nightingale, J. A., Hydroxypropyl methylcellulose acetate succinate-based spray-dried dispersions: an overview. *Mol. Pharm.* 2008, 5, 1003–1019.

Guzman, H.; Tawa, M.; Zhang, Z.; Ratanabanangkoon, P.; Shaw, P.; Gardner, C. R.; Chen, H.; Moreau, J. P.; Almarsson, O.; Remenar, J. F., Combined use of crystalline salt forms and precipitation inhibitors to improve oral absorption of celecoxib from solid oral formulations. *J. Pharm. Sci.* 2007, 96, 2686–2702.

He, Y.; Ho, C., Amorphous solid dispersions: utilization and challenges in drug discovery and development. *J. Pharm. Sci.* 2015, 104, 3237–3258.

von Helmholtz, R., Untersuchungen über Dämpfe und Nebel, Besonders über Solche von Lösungen. *Ann. Phys.* 1886, 263, 508–543.

Higashi, K.; Hayashi, H.; Yamamoto, K.; Moribe, K., The effect of drug and EUDRAGIT(R) S 100 miscibility in solid dispersion on the drug and polymer dissolution rate. *Int. J. Pharm.* 2015, 494, 9–16.

Higgins, J.; Cartwright, M. E.; Templeton, A. C., Progressing preclinical drug candidates: strategies on preclinical safety studies and the quest for adequate exposure. *Drug Discov. Today* 2012, 17, 828–836.

ICH M3(R2). 2009. ICH Guidance on Nonclinical Safety Studies for the Conduct of Human Clinical Trials and Marketing Authorization for Pharmaceuticals. Available at http://www.ich.org/fileadmin/Public_Web_Site/ICH_Products/Guidelines/Multidisciplinary/M3_R2/Step4/M3_R2__Guideline.pdf (accessed October 5, 2016).

Jantratid, E.; Dressman, J., Biorelevant dissolution media simulating conditions in the proximal human gastrointestinal tract: an update. *Dissolut. Technol.* 2009, 16, 21–25.

Kadajji, V.; Betageri, G., Water soluble polymers for pharmaceutical applications. *Polymers* 2011, 3, 1972–2009.

Kesisoglou, F.; Mitra, A., Crystalline nanosuspensions as potential toxicology and clinical oral formulations for BCS II/IV compounds. *AAPS J.* 2012, 14, 677–687.

Kesisoglou, F.; Panmai, S.; Wu, Y., Nanosizing-oral formulation development and biopharmaceutical evaluation. *Adv. Drug Deliv. Rev.* 2007, 59, 631–644.

Kola, I.; Landis, J., Can the pharmaceutical industry reduce attrition rates? *Nat. Rev. Drug Discov.* 2004, 3, 711–715.

Kramer, J. A.; Sagartz, J. E.; Morris, D. L., The application of discovery toxicology and pathology towards the design of safer pharmaceutical lead candidates. *Nat. Rev. Drug Discov.* 2007, 6, 636–649.

Kwong, E.; Higgins, J.; Templeton, A. C., Strategies for bringing drug delivery tools into discovery. *Int. J. Pharm.* 2011, 412, 1–7.

Leuner, C.; Dressman, J., Improving drug solubility for oral delivery using solid dispersions. *Eur. J. Pharm. Biopharm.* 2000, 50, 47–60.

Leung, D. H.; Lamberto, D. J.; Liu, L.; Kwong, E.; Nelson, T.; Rhodes, T.; Bak, A., A new and improved method for the preparation of drug nanosuspension formulations using acoustic mixing technology. *Int. J. Pharm.* 2014a, 473, 10–19.

Leung, D.; Nelson, T. D.; Rhodes, T. A.; Kwong, E. 2014b. Nano-suspension Process. US Patent US2A1 (October 26, 2012).

Li, P.; Zhao, L., Developing early formulations: practice and perspective. *Int. J. Pharm.* 2007, 341, 1–19.

Maas, J.; Kamm, W.; Hauck, G., An integrated early formulation strategy—from hit evaluation to preclinical candidate profiling. *Eur. J. Pharm. Biopharm.* 2007, 66, 1–10.

Manksy, P.; Dai, W.; Li, S.; Pollock-Dove, C.; Daehne, K.; Dong, L.; Eichenbaum, G., Screening method to identify preclinical liquid and semi-solid formulations for low solubility compounds: miniaturization and automation of solvent casting and dissolution testing. *J. Pharm. Sci.* 2007, 96, 1548–1563.

Merisko-Liversidge, E. M.; Liversidge, G. G., Drug nanoparticles: formulating poorly water-soluble compounds. *Toxicol. Pathol.* 2008, 36, 43–48.

Merisko-Liversidge, E. M.; Liversidge, G. G., Nanosizing for oral and parenteral drug delivery: a perspective on formulating poorly water-soluble compounds using wet media milling technology. *Adv. Drug Deliv. Rev.* 2011, 63, 427–440.

Merisko-Liversidge, E. M.; Liversidge, G. G.; Cooper, E. R., Nanosizing: a formulation approach for poorly water-soluble compounds. *Eur. J. Pharm. Sci.* 2003, 18, 113–120.

van den Mooter, G., The use of amorphous solid dispersions: a formulation strategy to overcome poor solubility and dissolution rate. *Drug Discov. Today* 2012, 9, e79–e85.

Moser, J. D.; Broyles, J.; Liu, L.; Miller, E.; Wang, M., Enhancing bioavailability of poorly soluble drugs using spray dried solid dispersions. Part I. *Am. Pharm. Rev.* 2008, 11, 68–71.

Muller, R. H.; Peters, K., Nanosuspensions for the formulation of poorly soluble drugs: I. Preparation by a size-reduction technique. *Int. J. Pharm.* 1998, 160, 229–237.

Muller, R. H.; Gohla, S.; Keck, C. M., State of the art of nanocrystals—special features, production, nanotechnology aspects and intracellular delivery. *Eur. J. Pharm. Sci.* 2011, 78, 1–9.

Munos, B., Lessons from 60 years of pharmaceutical innovation. *Nat. Rev. Drug Discov.* 2009, 8, 959–968.

Noyes, A. A.; Whitney, W. R., The rate of solution of solid substances in their own solutions. *J. Am. Chem. Soc.* 1897, 19, 930–934.

Oh, D.-M.; Curl, R. L.; Amidon, G. L., Estimating the fraction dose absorbed from suspensions of poorly soluble compounds in human: a mathematical model. *Pharm. Res.* 1993, 10, 264–270.

Oliver, I.; Shelukar, S.; Thompson, K. C., Nanomedicines in the treatment of emesis during chemotheraphy: focus on aprepitant. *Int. J. Nanomedicine* 2007, 2, 12–18.

Ormes, J. D.; Zhang, D.; Chen, A. M.; Hou, S.; Krueger, D.; Nelson, T.; Templeton, A., Design of experiments utilization to map the processing capabilities of a micro-spray dyer: particle design and throughput optimization in support of drug discovery. *Pharm. Dev. Technol.* 2013, 18, 121–129.

Paudel, A.; Geppi, M.; van den Mooter, G., Structural and dynamic properties of amorphous solid dispersions: the role of solid-state nuclear magnetic resonance spectroscopy and relaxometry. *J. Pharm. Sci.* 2014, 103, 2635–2662.

Peltonen, L.; Hirvonen, J., Pharmaceutical nanocrystals by nanomilling: critical process parameters, particle fracturing and stabilization methods. *J. Pharm. Pharmacol.* 2010, 62, 1569–1579.

PhRMA. 2012. Profile Pharmaceutical Industry. Available at http://www.phrma.org/sites/default/files/pdf/phrma_industry_profile.pdf (accessed October 5, 2016).

Pu, X.; Sun, J.; Li, M.; He, Z., Formulation of nanosuspensions as a new approach for the delivery of poorly soluble drugs. *Curr. Nanosci.* 2009, 5, 417–427.

Rabinow, B. E., Nanosuspensions in drug delivery. *Nat. Rev. Drug Discov.* 2004, 3, 785–796.

Raina, S.; Alonzo, D. E.; Zhang, G.; Gao, Y.; Taylor, L., Impact of polymers on crystallization and phase transition kinetics of amorphous nifedipine during dissolution in aqueous media. *Mol. Pharm.* 2014, 11, 3565–3576.

Rohrs, B. R., Biopharmaceutics modeling and the role of dose and formulation on oral exposure. In: *Optimizing the "Drug-Like" Properties of Leads in Drug Discovery*, Borchardt, R. T.; Kerns, E. H.; Hageman, M. J.; Thakker, D. R.; Stevens, J. L., Eds. Vol. IV of the Series Biotechnology: Pharmaceutical Aspects, pp. 151–166, Springer: New York, NY, 2006.

Scannell, J. W.; Blanckley, A.; Boldon, H.; Warrington, B., Diagnosing the decline in pharmaceutical R&D efficiency. *Nat. Rev. Drug Discov.* 2012, 11, 191–200.

Shah, A. K.; Agnihotri, S. A., Recent advances and novel strategies in pre-clinical formulation development: an overview. *J. Control. Release* 2011, 156, 281–296.

Shanbhag, A.; Rabel, S.; Nauka, E.; Casadevall, G.; Shivanand, P.; Eichenbaum, G.; Manksy, P., Method for screening of solid dispersion formulations of low-solubility compounds—miniaturization and automation of solvent casting and dissolution testing. *Int. J. Pharm.* 2008, 351, 209–218.

Shen, L. J.; Wu, F.-L., Nanomedicines in renal transplant rejection—focus on sirolimus. *Int. J. Nanomedicine* 2007, 2, 25–32.

Stegemann, S.; Leveiller, F.; Franchi, D.; de Jong, H.; Linden, H., When poor solubility becomes an issue: from early stage to proof of concept. *Eur. J. Pharm. Sci.* 2007, 31, 249–261.

Teja, S.; Patil, S.; Shete, G.; Patel, S.; Bansal, A., Drug-excipient behaviour in polymeric amorphous solid dispersions. *J. Excip. Food Chem.* 2013, 4, 70–94.

Vandercruysa, R.; Peeters, J.; Verreck, G.; Brewster, M. E., Use of a screening method to determine excipients which optimize the extent and stability of supersaturated drug solutions and application of this system to solid formulation design. *Int. J. Pharm.* 2007, 342, 168–175.

Vasconcelos, T.; Sarmento, B.; Costa, P., Solid dispersions as strategy to improve oral bioavailability of poor water soluble drugs. *Drug Discov. Today* 2007, 12, 1068–1075.

Venkatesh, S.; Lipper, R. A., Role of the development scientist in compound lead selection and optimization. *J. Pharm. Sci.* 2000, 89, 145–154.

Warren, D. B.; Benameur, H.; Porter, C. J. H.; Pouton, C. W., Using polymeric precipitation inhibitors to improve the absorption of poorly water-soluble drugs: a mechanistic basis for utility. *J. Drug Target.* 2010, 18, 704–731.

Wu, Y.; Loper, A.; Landis, E.; Hettrick, L.; Novak, L.; Lynn, K.; Chen, C.; Thompson, K.; Higgins, R.; Batra, U.; Shelukar, S.; Kwei, W.; Storey, D., The role of biopharmaceutics in the development of a clinical nanoparticle formulation of MK-0869: a Beagle dog model predicts improved bioavailability and diminished food effect on absorption in human. *Int. J. Pharm.* 2004, 285, 135–146.

Wu, L.; Zhang, J.; Watanabe, W., Physical and chemical stability of drug nanoparticles. *Adv. Drug Deliv. Rev.* 2011, 63, 456–469.

Xu, S.; Dai, W.-G., Drug precipitation inhibitors in supersaturable formulations. *Int. J. Pharm.* 2013, 453, 36–43.

7

Formulation, Analytical, and Regulatory Strategies for First-in-Human Clinical Trials

Lorenzo Capretto, Gerard Byrne*, Sarah Trenfield, Lee Dowden and Steven Booth*

Merck Sharpe & Dohme, Hoddesdon, Hertfordshire, UK

---

* These authors contributed equally to this work.

Oral Formulation Roadmap from Early Drug Discovery to Development,
First Edition. Edited by Elizabeth Kwong.
© 2017 John Wiley & Sons, Inc. Published 2017 by John Wiley & Sons, Inc.

7.1 Introduction

First-in-human (FIH) clinical trials are an essential step in the drug development process and represent the first application of an investigational medicinal product (IMP) in humans. Primarily, these studies aim to understand the investigated drug's safety, tolerability, and pharmacokinetic behavior to advise a suitable dosing regimen for later Phase I and II clinical trials. However, these studies often involve an initial investigation of the efficacy of the IMP through the study of biomarkers in both volunteers and patients. Figure 7.1 outlines the main aims and the typical number of volunteers (or patients) recruited within each major stage of the drug development process.

As the data collected from these studies paves the way for future clinical decisions, it is critical to design the trial appropriately to ensure both data reliability and patient safety.

Effectively planning and designing an FIH trial will ensure the safety of study volunteers, as well as maximizing efficiency and reducing resource investment in the clinical trial process. This chapter will detail the major steps of conducting an FIH clinical trial. Initially, the chapter will discuss the planning and execution of the FIH study; subsequently, it will introduce the formulation and

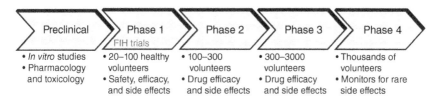

Figure 7.1 Major stages of the drug development process (FDA, 2015).

analytical development activities necessary to supply the drug product (DP) for the clinical trial; finally, it will examine the preparation of the regulatory submission for the clinical trial application.

7.2 Planning and Executing the FIH Trial

There are a number of planning steps required before taking a new chemical entity (NCE) into clinical trials, including study protocol and investigator brochure development, and regulatory filing submissions. The overall process from NCE selection to dosing in humans is depicted in Figure 7.2.

The study protocol should describe how the trial will be conducted by covering background information, study rationale and objective(s), trial design, methodology, data management, and statistical considerations and will ensure the safety of trial subjects and integrity of data collected. The International Conference on Harmonization (ICH) Good Clinical Practice E6(R1) Guidance details the necessary sections that should be contained in a study protocol and is a good reference source to follow (ICH, 1996a).

A financial management plan should also be devised, outlining the trial budget alongside the anticipated cost of research items, trial site(s), and staff. Fees from any outsourced services such as contract research organizations (CROs) should also be considered. CROs have expertise in conducting clinical trials and will often assist in site selection, patient recruitment, clinical monitoring, and data collection and management. Outsourcing activities to these companies can be useful if there are no in-house capabilities for these tasks or experience in clinical trial conduct is limited.

In the United Kingdom, a clinical trial authorization (CTA) is required from the UK's regulatory agency, the Medicines and Healthcare products Regulatory Agency (MHRA), before dosing in humans can commence. In the United States, an Investigational New Drug (IND) designation must be provided from the Food and Drug Administration (FDA). In both cases, nonclinical data (e.g., toxicology studies) must be submitted, along with the trial protocol, Investigator's Brochure, and chemistry, manufacturing, and controls (CMC) information, among others (FDA, 2014; MHRA, 2014a). An independent research ethics committee (IEC), also known as an institutional review board

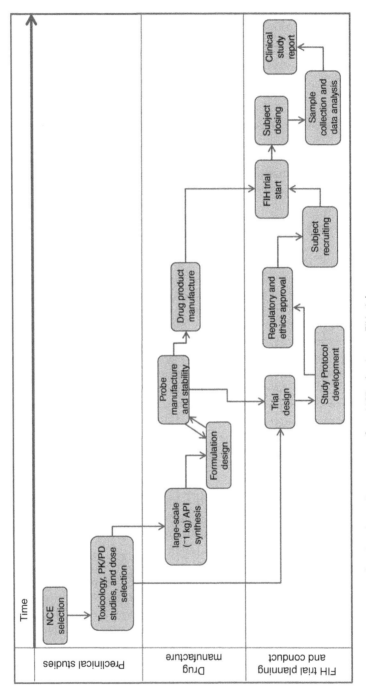

Figure 7.2 Route map showing the major processes from NCE selection to FIH trial.

(IRB), must also review study material. An IEC acts to protect the rights, safety, and well-being of trial subjects and aims to provide public assurance of that protection. This will include the review of the trial site and investigator suitability and the trial protocol and methods to obtain informed consent. The legal requirements and IEC composition can vary from country to country, but every IEC should follow the ICH's Good Clinical Practice E6(R1) Guidance (ICH, 1996a).

Consideration of subject eligibility, subject numbers, trial design, and data analysis capabilities should be evaluated before the study protocol is finalized and submitted for regulatory and ethical approval. The following sections will discuss important considerations for each of these.

7.2.1 Study Population

Within an FIH trial, an appropriate sample size of a suitable subject population is required; the aim is to produce reliable data without dosing an excessive number of volunteers. The number of subjects is determined by the nature of the drug, the trial design, and the overall study aims. In practice, the majority of non-oncology FIH clinical trials are performed in non-smoking, healthy volunteers aged 18–55 years, as they produce the "cleanest" data and act as a good model of drug behavior in man (Dresser, 2009).

However, in certain situations the use of patients might be favored. For example, in oncology trials, patients are frequently recruited, as it would be unethical to deny a patient treatment and undesirable to expose healthy volunteers to a toxic drug (e.g., cytotoxics) (Dresser, 2009). In later Phase I trials, there may also be parts to the study to evaluate a drug's pharmacodynamics, which might only be feasible in patient populations, for example, if they exhibit the drug target as opposed to healthy subjects.

However, enlisting patients can provide additional complications. Subject enrolment times will often be increased due to a greater difficulty of identifying suitable individuals. Furthermore, patients might already be taking concomitant medication to manage their condition that could have the potential to interact with the investigated drug, potentially altering its absorption, distribution, metabolism, and excretion (ADME) profile. This could produce unreliable results and also affect the choice of dose for future trials. If the risk of interaction is high, it would be advisable to carry out drug–drug interaction (DDI) studies before starting the multiple ascending dose (MAD) studies, which will inform the dose to bring forward to later phase trials.

Before commencing an FIH trial, research teams are required to screen subjects to find those that are eligible for participation in the study. Inclusion and exclusion criteria must be written as part of the study protocol to advise subject recruitment. Typical exclusion conditions include pregnant women and children, due to the unknown drug effects on the unborn fetus and children, respectively. Historically, women of childbearing potential (WOCBP) have been excluded from early phase clinical trials because of this reason. The FDA have since adapted their guidance and now actively encourage the participation of WOCBP in early phase studies, as long as appropriate safety precautions are in place, for example, use of highly effective birth control methods and pregnancy testing (FDA, 1993; ICH, 2009a).

It is also important to consider collecting pharmacokinetic information in other subpopulations, such as the elderly and ethnic subgroups. For example, by including subjects of different ethnicities in FIH or Phase I studies, variations in response between ethnicities can be detected early on. This poses a significant advantage of reducing development times downstream by gaining an understanding of whether the drug is ethnically sensitive early on, hence enabling sponsors to plan accordingly for later phase studies. In this regard, it is important to note that before starting a Phase II study in Chinese and Japanese populations, it is required to run a Phase I study on similar ethnic subgroups. The ICH E5(R1) Guidance provides information on ethnic subpopulations in trials and how to conduct supplementary ethno-bridging trials (ICH, 1998).

To calculate a suitable sample size, a balance must be struck between data reliability and subject recruitment. Consideration of the nature of the drug (i.e., the expected response or toxicity profile), the risk of error, and the level of significance to achieve the desired "power of the study" is required (Sakpal, 2010). This power estimates the ability to measure a true difference between the intervention and control groups. Typically, a small number of 20–80 subjects are recruited for FIH trials (Buoen *et al.*, 2005). An internal literature review found that there was an average of 38 subjects enrolled onto non-oncology FIH trials between 1995 and 2014 (Merck Sharp & Dohme Ltd, Unpublished research).

The sample size required can vary according to study design; crossover designs have greater statistical efficiency, thus requiring fewer study participants than a parallel design. An internal literature review showed that an average of 44 subjects were recruited onto a parallel single dose design ($n = 59$) compared with an average of 29 subjects recruited onto a crossover single dose design ($n = 19$) (Merck Sharp & Dohme Ltd, Unpublished research). In practice, companies will often choose a trial design that will ensure patient safety and data reliability with a minimal number of subjects.

7.2.2 Site and Investigator Selection

Choosing where to conduct a clinical trial requires the review of a number of factors to ensure the trial demands are met. These include the anticipated rate of participant recruitment, cost of trial site, site's experience in conducting clinical trials, geographical location, and ease of access to emergency and supportive facilities (e.g., resuscitation, antidotes, intensive care units, etc.).

Furthermore, if a formulation is chosen that requires manufacturing before dosing (e.g., the preparation of an oral suspension from a powder with a suspending agent), the on-site preparation facilities should be assessed for suitability. In addition, if the study is carried out in the European Union (EU), the site must have a manufacturing and assembly license (such as a manufacturer/importer license (MIA) that is issued by the MHRA in the United Kingdom) in order to undertake the formulation manufacture (MHRA, 2014b). The site must also have timely access to a qualified person (QP) who will certify the release and use of the prepared medicine in humans.

There are schemes to aid sponsors in site selection; in the United Kingdom, a voluntary Phase I supplementary accreditation involves a thorough inspection by the MHRA to ensure that the site is appropriate for clinical trial conduct (MHRA, 2014a). As far as the authors are aware, the FDA has not released such guidance, and therefore the sponsor has ultimate responsibility to assess the site's suitability within the United States. Outsourcing trial-related duties to CROs will enable the selection of a trial site and study investigator that have been deemed reputable by their organization.

As well as to choose a suitable site, it is important to select an appropriate study investigator. Within the United States, EU, and further afield, study investigators must hold sufficient evidence to prove that they can competently and safely perform the trial. This was written in the *Declaration of Helsinki*, which is a set of ethical principles for medical research on human subjects (World Medical Association, 1964). Within the United Kingdom, investigators must hold a postgraduate qualification relating to pharmacology, for example, a diploma of clinical pharmacology (MHRA, 2014c). Within the United States, the FDA do not explicitly define the qualifications that the investigator must hold but do state that they should have appropriate expertise (FDA, 2010).

Before commencing a clinical trial, an Investigator's Brochure must be developed to provide information to study investigators and other trial staff. This is a document that compiles all clinical and nonclinical data about the IMP that is relevant to its dosing in humans. It should explain the rationale behind the choice of dose selection, dose frequency, method of administration, and safety monitoring procedures. Furthermore, it should cover the drug properties and formulation, nonclinical study results (pharmacokinetics, metabolism, and toxicology in animals), and anticipated effects in humans (pharmacokinetics, safety, and efficacy). The ICH Good Clinical Practice E6(R1) Guidance provides a detailed explanation for each section (ICH, 1996a).

7.2.3 FIH Trial Design

Appropriately designing a clinical trial will ensure patient safety as well as data reliability. A number of areas should be included within the study protocol, such as dose determination, dose escalation strategies, and blinding techniques, each of which will be discussed in the succeeding sections. It should be noted that this section mainly covers non-oncology FIH trials; Section 7.2.4 will discuss this information in relation to oncology-specific trials.

7.2.3.1 Initial Dose Determination

The initial dose used in FIH clinical trials should ensure a balance between subject safety and efficiency of dose escalation to reach target concentrations. The European Medicines Agency (EMA) and the FDA have released guidance to advise a suitable dose selection, the content of which is summarized in the succeeding text (EMA, 2007; FDA, 2005).

Typically, the no observed adverse effect level (NOAEL) is calculated during preclinical toxicology studies; this is defined as the highest dose administered to a relevant animal species that does not produce a statistically or biologically significant adverse effect compared with a control (FDA, 2005). This value is adjusted to account for any interspecies differences (e.g., body surface area) and anticipated human exposure, which calculates a human equivalent dose (HED). The species that generates the lowest HED is considered the most sensitive and the most appropriate species. Therefore, this value is used and adjusted by a greater than or equal to 10-fold safety factor to advise the maximum recommended starting dose (MRSD) (Muller *et al.*, 2009). These adjustments aim to provide confidence that the initial dose in man is low enough to prevent adverse effects (FDA, 2005).

While NOAEL has traditionally been used within FIH trials, the EMA has now placed a greater emphasis on moving away from this model for those drugs with higher risk, for example, those that act on an ubiquitously expressed receptors or that are involved with cytokine release (e.g., CD4 or CD28 agonists) (EMA, 2007). It is recommended to use the minimum anticipated biological effect level (MABEL) approach for those drugs, which calculates an anticipated dose required to produce a minimal biological effect in humans. As defined by the EMA (2007), it should take into account the following data and implement it into a pharmacokinetic/pharmacodynamic model, where possible:

- *In vitro* target binding and receptor occupancy data in human cells and relevant animal species
- *In vitro* concentration–response curves from human cells and relevant animal species
- *In vivo* dose/exposure–response curves from a relevant animal species
- Exposure of pharmacological doses to relevant animal species

A safety factor should be applied to the MABEL value taking into account the differences between human and animal physiology, as well as drug potency, method of action, and other factors such as novelty and shape of the dose–response curve. This will then produce the MRSD for FIH clinical trials (EMA, 2007; Muller *et al.*, 2009).

7.2.3.2 Dose Increment and Escalation Strategies

Once an initial dose and dose range has been determined, implementation into a dose escalation strategy by defined dose increments should be undertaken. There are a number of dose increment designs that fall into four main categories (Figure 7.3), the choice of which should be determined on a case-by-case basis.

Linear dose designs involve a predefined fixed dose increment escalation; for example, the dose increases by 5 mg upon each dose level: 5 mg → 10 mg → 15 mg and so on. A logarithmic design involves the relative dose increment being fixed; for example, 100% of the previous dose is administered upon each dose level: 5 mg → 10 mg → 20 mg → 40 mg. Both of these designs were frequently used in non-oncology FIH trials with 12% of studies using the linear design and 22% using the logarithmic design (Buoen *et al.*, 2005).

The modified Fibonacci design has been historically used within oncology-specific FIH trials; however it can sometimes be applied to non-oncology trials. The true Fibonacci sequence follows the recurrence relationship rule:

$$F_n = F_{n-1} + F_{n-2}; \quad F_0 = 0, \; F_1 = 1$$

In other words, the next number in the sequence is the sum of the two predecessors; 0, 1, 1, 2, 3, 5, 8, 13, 21, ... Ratios can be calculated by F_n/F_{n-1} (for $n \geq 3$) to produce values of 2.00, 1.50, 1.67, 1.60, 1.63, 1.62 and so on. Within FIH trials, this sequence is modified to follow predefined incremental dose ratios. An example of the commonly used ratio sequence is 2.00, 1.67, 1.50, and 1.33, with all successive dose levels using 1.33 (Rogatko *et al.*, 2007).

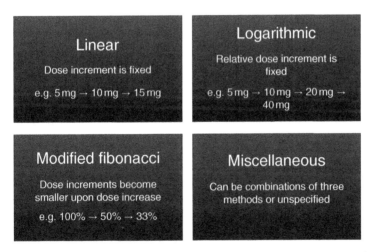

Figure 7.3 The four main designs utilized for dose escalation in FIH clinical trials between 1995 and 2004. Source: Buoen *et al.* (2005). Reproduced with permission of John Wiley & Sons.

However, it is apparent that not all published FIH trials always follow these set regimens; an internal literature review showed that the range of dose increments varied between 15 and 700% (Merck Sharp & Dohme Ltd, Unpublished research). This supports Buoen *et al.*'s (2005) research, which found that 64% of FIH and Phase I trials did not follow a set dose escalation regime, either being a combination of linear, logarithmic, and Fibonacci designs or being "arbitrarily chosen."

The choice of dose increment should depend on the nature of the drug and preclinical safety data. For example, smaller dose increments should be used for those IMPs that exhibit a high risk of toxicity. It is worth noting that the dose increments can be altered during the trial, as long as flexible wording within the study protocol permits this, based on the results of previous dosing and safety.

7.2.3.3   Blinding

Blinding is the process of concealing research design aspects from subjects, study investigators, or data collectors (Page and Persch, 2013). Various blinding techniques are often used in clinical trials to reduce the risk of bias introduced by expectations and feelings relating to drug treatment. Subjects might be blinded to gain reliable information about drug toxicity and tolerability. Staff can also be blinded to prevent any subconscious or (un)intentional bias relating to subject treatment and resulting data analysis, which could affect future dosing decisions.

Figure 7.4 shows the most commonly used strategies, which include open-label, single-blind, and double-blind designs. Double blinding has the advantage of reducing the risk of bias within results and is often the most favored strategy.

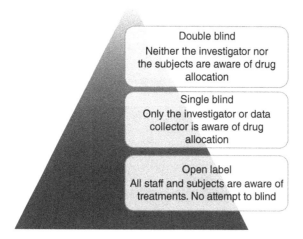

Figure 7.4 A pyramid diagram to show the three types of blinding designs. Source: Page and Persch (2013). Adapted with permission of American Journal of Occupational Therapy.

Methods of blinding could involve manufacturing a placebo with the same characteristics as the active and using coding protocols to anonymize the drug/placebo name on the packaging (Page and Persch, 2013).

Blinding is common within non-oncology FIH trials, with an internal literature review showing that 85% of studies between 1995 and 2014 were double blinded. Conversely, 98% of oncology-specific FIH trials were found to use an open-label strategy and less likely to undertake patient randomization (Merck Sharp & Dohme Ltd, Unpublished research).

Once the method of dose escalation and blinding has been confirmed, this can then be implemented into a number of trial designs. The general approach is to randomize subjects into cohorts and undertake sequential dose escalations, using a placebo as a control throughout. The most common trial designs for non-oncology FIH trials include parallel and crossover designs.

7.2.3.4 Single Ascending Dose (SAD) Studies

Parallel single dose designs involve every study volunteer in each cohort receiving only one administration of one dose. After a predefined time period and after a safety evaluation has been performed, involving detection of adverse events and tolerability, the second cohort will be given the next dose level as shown in Table 7.1 (Association of the British Pharmaceutical Industry, 2011).

This design would be most suitable for those drugs that exhibit a high risk of toxicity upon multiple administrations and for those with long half-lives that would be unacceptable within a crossover design, without a sufficient washout period. However, a disadvantage is the large number of volunteers needed compared with other designs, as well as the absence of understanding around intra-patient variability.

Crossover single dose designs consist of alternating dose escalations between different cohorts, with each subject acting as their own control. This enables investigators to gain additional data about intra-subject variability as well

Table 7.1 Parallel single ascending dose design with a linear dose escalation. Safety column represents safety evaluation prior to subsequent dosing.

	Time 1	Time 2	Time 3	Time 4	Time 5
Cohort 5					25 mg *Safety*
Cohort 4				20 mg *Safety*	
Cohort 3			15 mg *Safety*		
Cohort 2		10 mg *Safety*			
Cohort 1	5 mg *Safety*				

as differences between subjects (Association of the British Pharmaceutical Industry, 2011). Placebo administration can be randomized throughout each cohort or be attributed to specific subject in each cohort.

There are two main crossover study designs: standard and interlocking. The standard design involves each subject in a cohort receiving a predefined number of increasing doses or a single dose of placebo. After a safety evaluation, the next cohort will receive subsequent dose increases (Table 7.2).

In the interlocking crossover design, each cohort receives every other (or every third) dose escalation (Table 7.3). This provides the advantage of allowing longer washout periods between each dose, particularly important for those drugs with long half-lives, and of preventing confounding results by previous pharmacokinetic or pharmacodynamic effects. Furthermore, the overall trial length is often reduced compared with parallel designs. However, the increase in participation time required for the volunteers might increase the likelihood of dropouts from the study.

Both designs distribute the total quantity of subjects into cohorts of around 6–12 volunteers. An internal literature review found that the most common ratio of the number of subjects receiving the active drug compared with a placebo is 6:2, with the second most common being 8:2 (Merck Sharp & Dohme Ltd, Unpublished research). However, variations of this ratio do exist, including 4:3, 5:1, and 5:2. The ratio used should provide confidence to the investigator that any observed drug effects are real.

Table 7.2 Standard crossover design with a linear dose escalation. Safety column represents safety evaluation prior to subsequent dosing.

	Time 1		Time 2		Time 3		Time 4		Time 5		Time 6	
Cohort 2							20 mg	Safety	25 mg	Safety	30 mg	Safety
Cohort 1	5 mg	Safety	10 mg	Safety	15 mg	Safety						

Table 7.3 Interlocking cohort's crossover design with a linear dose escalation. Safety column represents safety evaluation prior to subsequent dosing.

	Time 1		Time 2		Time 3		Time 4		Time 5		Time 6	
Cohort 2			10 mg	Safety			20 mg	Safety			30 mg	Safety
Cohort 1	5 mg	Safety			15 mg	Safety			25 mg	Safety		

7.2.3.5   Multiple Ascending Dose (MAD) Studies

Parallel or crossover multiple dose designs are typically conducted following the completion of a single ascending dose (SAD) study, extending into the Phase I remit. The predominant aim is to ascertain the maximum tolerated multiple dose level and to assess the drug's pharmacokinetic behavior at steady state. Within a parallel design, subjects are administered with repeated doses at the same dose level, and, after a safety evaluation, the next cohort are administered with the next dose level and so on (Table 7.4).

The parallel multiple dose design is particularly useful for assessing the effects of repeated dose administration and to detect toxic accumulation. However, as with the parallel single dose design, there is no consideration of intra-subject variability; the use of a crossover multiple dose design could be considered in this situation.

7.2.3.6   Multi-Part Flexible Trial Designs

In recent years, a more adaptive approach to clinical trial design has been developed, termed multipart flexible trial designs. These designs differ in that they can include SAD, FIH, MAD, and proof-of-concept studies within a single trial under a single adaptive protocol.

This flexible protocol can allow the use of real-time data to guide decisions about dosing and formulation selection throughout the trial, as well as assessing other factors such as food effects and proof of concept. For example, a study performed in 2015 used a flexible design to assess the biopharmaceutical performance of a solution versus a capsule formulation in an initial SAD part to the study (Millington *et al.*, 2015). They also included the option to assess a second type of capsule, food effects, and different dosing regimens in this section. Data collected from the SAD was used to select the characteristics of testing in the MAD part (e.g., starting dose, formulation selection, and fasted-/fed-state dosing). Other studies have included testing in both healthy volunteers and patients within the same study, as well as assessing drug pharmacodynamics in a proof-of-concept section to the study (Patrick *et al.*, 2012).

Advantageously, these adaptive designs can facilitate a more rapid progression from testing in healthy volunteers to patients and can avoid conducting a stand-alone bioavailability study when transitioning to a different formulation downstream. This can reduce early phase resource investment and trial duration.

7.2.4   Oncology-Specific FIH Trial Designs

In contrast to general FIH studies, oncology trials are often unblinded, use a MAD study design, and do not use a placebo as a control. Furthermore, those trials using cytotoxic agents are often only available to patients with advanced cancers or for those who have had little success with prior treatments due to

Table 7.4 Parallel multiple dose design with a linear dose escalation. Placebo allocation can be randomized throughout. Safety column represents safety evaluation prior to subsequent dosing.

	Time 1		Time 2		Time 3		Time 4		Time 5		Time 6		Time 7		Time 8		Time 9	
Cohort 3													15 mg	Safety	15 mg	Safety	15 mg	Safety
Cohort 2							10 mg	Safety	10 mg	Safety	10 mg	Safety	10 mg	Safety	10 mg	Safety	10 mg	Safety
Cohort 1	5 mg	Safety	5 mg	Safety	5 mg	Safety	5 mg	Safety	5 mg	Safety	5 mg	Safety	5 mg	Safety	5 mg	Safety	5 mg	Safety

their relatively nonselective effects on tumor versus normal cells (Shivaani *et al.*, 2006). As previously discussed, cytotoxic agents can exhibit substantial toxicity, and therefore testing in healthy volunteers is often avoided. Healthy volunteers can be used, however, when testing more selective, molecularly targeted agents to gain an initial idea around drug pharmacokinetics, toxicity, and safety.

The starting dose for FIH oncology trials should be based on preclinical animal toxicology studies, as recommended under ICH Topic S9 Guidance (ICH, 2009b). This guidance does not explicitly define how to determine a starting dose but does provide information about typically used dose calculations. For example, for small molecule cytotoxic agents, the initial dose is commonly determined by using 1/10th of the dose that caused severe toxicity in rodents, known as the severe toxicity dose (STD_{10}). The STD_{10} can be used as long as this dose does not cause severe or irreversible reactions when tested in non-rodent species. If severe toxicities are observed at the STD_{10} in non-rodents or if the non-rodent is known to be a more appropriate model, then the dose that is 1/6th of the highest non-severely toxic dose can be used. To account for interspecies differences, consideration of physiological differences and scaling according to body surface area must be conducted.

To advise dosing in later phases, the maximum tolerated dose (MTD) is calculated during oncology-specific FIH trials; this is the dose where the probability of a medically unacceptable dose-limiting toxicity (DLT) is equal to a predefined target value, which is determined by the nature of the drug and the associated DLT (Yin and Yuan, 2009). For example, if it is an irreversible, fatal toxic reaction, then the target value will be low. The MTD is commonly calculated in oncology due to the widely accepted theory that higher doses of cytotoxic drugs enhance therapeutic benefit, but it must be considered that higher doses can also cause substantial drug toxicity.

The design strategies used for oncology trials are broadly classed as rule-based or model-based designs (Tourneau *et al.*, 2009). Rule-based designs assume little prior knowledge about the dose–toxicity curve and enable rapid escalation and de-escalation of doses; the main strategies include the 3 + 3 design and the accelerated titration design (ATD). Model-based designs differ in that they utilize a statistical model to calculate a dose based on the probability of a DLT occurring (Tourneau *et al.*, 2009).

7.2.4.1 Rule-Based Designs

The 3 + 3 design involves using cohorts each containing three subjects; the first is treated at a predefined initial dose, and subsequent cohorts are administered with increasing doses. The continuation of dosing depends heavily upon any toxicity observed by the previous cohort of subjects. For example, if no patients experience a DLT, the next dose will be administered to the next cohort. However, if one subject experiences toxicity, then three more patients will need

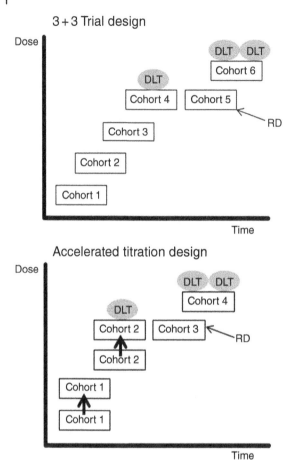

Figure 7.5 3 + 3 trial design. Each cohort contains three subjects and dose is escalated until two DLTs are observed. *RD* recommended dose for later phase clinical trials. *ATD* accelerated titration design. Upward arrows represent intra-patient dose escalation.

to be treated at the same dose level (Figure 7.5). This dose escalation regime continues until two or more patients experience toxicity at the same dose level.

Variations of this design have been created and are less commonly used, such as the 2 + 4 design and 3 + 3 + 3 design (Tourneau *et al.*, 2009). A major disadvantage of the 3 + 3 design is that many dose escalation steps are needed to reach the MTD, resulting in a large number of patients being treated at potentially subtherapeutic doses. Consequently, the ATD was devised by Simon *et al.* (1997), which holds similarities to the 3 + 3 design in that there are sequential dose escalations and toxicity evaluations (Figure 7.5). However, there is usually only one patient per cohort, which reduces the total number of patients required compared with the 3 + 3 design. When a DLT is observed, the ATD often reverts to the 3 + 3 design.

The dosing of individual patients at more than one dose level can be used in oncology FIH trials, which is termed intra-patient dose escalation, shown in Figure 7.5; this technique takes into account intra-patient variability and also prevents subjects from being treated at subtherapeutic doses. Theoretically, this should enable studies to recruit fewer subjects. However, an internal literature review found the converse to be true; an average of 47 patients were recruited for those studies that used intra-patient dose escalation as opposed to 33 for those that did not (Merck Sharp & Dohme Ltd, Unpublished research).

7.2.4.2   Model-Based Designs

Model-based designs use a statistical model to calculate a dose based on the probability of a DLT occurring, for example, using Bayesian models (Yin and Yuan, 2009). The ongoing safety data collected throughout the trial can be inputted into these models to determine a safe dose in the next patient, as long as flexible wording in the study protocol allows this. Types of model-based designs include the continual reassessment method (CRM) and escalation with overdose control (EWOC).

The CRM, founded by O'Quigley and colleagues in 1990, is based upon patients being treated with a dose closest to the MTD and has the advantage of treating patients at higher, more effective dose levels (O'Quigley *et al.*, 1990). However, due to concerns that the original CRM exposes subjects to high toxic doses, modifications have been described in the literature, including the development of EWOC (Babb *et al.*, 1998). The EWOC method is an adapted Bayesian model with a similar structure to the CRM but with additional safety controls in place. Data is continually collected after dosing to calculate the probability of exceeding the MTD in future dose escalations.

7.2.5   Other Considerations

Although the main objective of an FIH trial is to assess drug tolerability, pharmacokinetics, and safety, there may also be other aspects that are worth exploring as a separate part to the study. Examples include assessing the effects of food or the formulation type on the drug pharmacokinetics or evaluating DDI.

It has been long known that food can significantly impair oral drug bioavailability by delaying gastric emptying, changing luminal pH, or even physically or chemically interacting with the drug substance (DS) or its formulation. To evaluate the impact of food on a drug's biopharmaceutical performance, food-effect bioavailability studies can be conducted. These studies assess the effect of food on drug absorption in subjects that are dosed in the fasted state compared with subjects who have consumed a high-fat and high-calorie meal (FDA, 2002). This can advise future recommendations when dosing in subjects in later phases.

If the IMP is predicted to be used alongside another regular therapy, a DDI study could be undertaken as a separate part to trial. This could provide information around the drug's safety profile, which could be used to guide any inclusion/exclusion criteria for later phase trials. For example, for the auto-immune condition rheumatoid arthritis, methotrexate is considered to be an "anchor" drug for patients. As such, a published FIH trial conducted an additional separate part to the study assessing whether the IMP and methotrexate interacted (Namour *et al.*, 2012).

If there is a difficulty in predicting human pharmacokinetics for the IMP and if this is the primary factor for driving a go or no go clinical decision, an exploratory human microdose study can be undertaken, which is supported in the ICH M3(R2) Guidance (ICH, 2009a). This design typically employs administering a subtherapeutic dose (<1/100th of a pharmacologically active dose), up to a maximum of 100 μg. This reduces the need for extensive preclinical toxicology studies and offers the advantage of eliminating drug candidates that exhibit poor pharmacokinetics *in vivo* early on. Furthermore determining the effect that different formulations (e.g., solution vs. tablet) or active pharmaceutical ingredient (API) properties on a drug's pharmacokinetic profile is sometimes undertaken as a separate part to the study, termed bioequivalence (BE) studies. These will be discussed in Section 7.3.7.

7.2.6 Sample Acquisition and Data Analysis

Once regulatory and ethical approval has been confirmed and the trial has been effectively designed, dosing subjects with the IMP can commence. It is important to have an appropriate management plan in place, including timings that subjects will be dosed, safety evaluations, sample acquisition (e.g., blood samples, urine collection), and pharmacokinetic/pharmacodynamic analysis.

The common pharmacokinetic parameters that are reviewed in SAD studies include the maximum plasma concentration (C_{max}), the time to maximum plasma concentration (T_{max}), and the total human drug exposure by calculating the area under the curve (AUC) (Figure 7.6). For MAD studies, these parameters at steady state can be assessed, which occurs when the amount of drug dosed is equal to the amount of drug being eliminated. This information can be collected between dosing of subjects, and the next dose level can be adjusted, assuming that the study protocol allows this (e.g., by using flexible wording).

If in-house analysis capabilities are limited, CROs often have expertise in data analysis and can arrange sample collection, perform data analysis, and report data back to the sponsor.

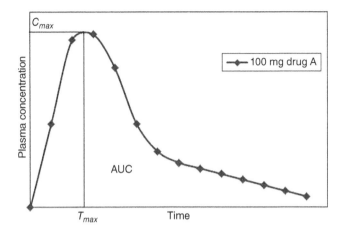

Figure 7.6 An example concentration–time graph showing C_{max}, T_{max}, and AUC.

7.2.7 Clinical Trial Completion

At the end of the FIH trial, sponsors must notify relevant authorities (e.g., MHRA in the United Kingdom, FDA in the United States) and ethics committees. The definition of the end of the trial should be included in the initial study protocol and is recommended in most cases to be the date of visit of the last patient to be dosed in the trial, but it can also represent the final data capture, especially if monitoring upon follow-up is required.

The sponsor must also provide a clinical trial summary report for review by regulatory authorities and ethics committees within 1 year of the end of the trial for non-pediatric trials within the EU. The report must provide all methods and results, including adverse events, pharmacokinetic/pharmacodynamic information, outcome measure(s), and food effects, among others, from the clinical trial. The format of the study report should follow ICH E3 Guidance, Structure and Content of Clinical Study Reports (ICH, 1995).

7.3 Formulation Development

Small and large pharmaceutical companies are facing unprecedented pressure to reduce cost and accelerate drug development candidates to market while increasing the number of successful development programs (Allison, 2012). To achieve this goal, on one hand, there is the need to offer fast and inexpensive ways to enter the clinical trial phase with the minimal amount of resources to drive the pipeline productivity and meet financial constraints. On the other hand, as the cost and timeline of late stage clinical trials spiral, drug developers are required to answer key development questions in the early phase of clinical development to maximize the chance of rapid market access.

Pharmaceutical development for DPs to be used during FIH studies can be approached using two different formulation strategies: a *commercial* paradigm favors FIH formulations that are similar to the final marketable product; conversely, a *fit-for-purpose* (FFP) paradigm relies on the use of the simplest possible formulation tailored for the need of early phase clinical objectives and proof-of-concept studies. Typical FFP formulations include *API in bottle* where neat API is dispensed in a bottle or vials for later reconstitution before dosing, *API in capsule* where neat API is dispensed directly in a capsule, and *ready-to-use solutions* or *suspensions* where the API is formulated in a liquid media. There are also cases in which more developed formulations are used in which simple blends are formulated as solid dosage forms like *capsules* and *tablets*.

7.3.1 Fit-for-Purpose versus Commercial Approach

As discussed in the previous sections, the main objective of an FIH study is to assess the pharmacokinetics and tolerability of a new compound to determine the dose range to be used in the later stage of clinical evaluation. At such an early stage, the pharmacokinetics in human is often unknown, and there is a need for a large dosing flexibility over a wide dose range. This need might be associated with a limited availability of API for formulation and analytical development activities.

It is well known that the rate of failure is high during early phase drug development. In 2014, the likelihood of an FDA approval for an NCE in Phase I was only 10% (Hay *et al.*, 2014), which questions the need for extensive resource investment at the early FIH trial stage. The ideal drug delivery method for FIH trials would often be a simple, cost-effective formulation to support the earliest possible dosing in man.

Favorably, the use of FFP formulations can accelerate the drug into FIH studies while limiting the use of API and other resources. This permits the rapid screening of several potential preclinical compounds to counteract the high failure rate of early phase clinical development.

In an FFP strategy, the API is typically dosed as a neat component or with a limited number of additional excipients. Examples of excipients that are often avoided in FFP formulations are those required for industrial scale production (i.e., glidant and lubricant) and those required for commercial image development (i.e., coating agent and colors). This simplification not only greatly reduces development resources but also reduces the need for extensive process qualification activities and excipient compatibilities studies. By using an FFP strategy, companies can concentrate on the basic aspects of the DP qualification, accepting gaps that will be addressed in later stage development. For example, FFP formulations might not undergo extensive long-term and solid-state stability or full polymorphic profiling evaluation.

From an economical perspective, using an FFP formulation strategy is driven by the desire to reduce investment for a new compound prior to the demonstration of its clinical proof of concept. This allows for a better rationalization of resources that could be efficiently diverted toward those programs with less favorable physiochemical attributes (e.g., low solubility) and potential blockbusters.

The FFP approach offers an appealing alternative to enter early into the clinical trial phase; however scientists should be mindful of the disadvantages of developing a formulation that would be unusable beyond the FIH study. Generally speaking, using an FFP approach may ultimately result in greater development costs and time. In fact, the FFP approach defers the need for extensive product development and process optimization but not eliminate it. As a result, the FFP development resources will be additive to the total resources used during the full development journey. For example, entering the clinical trial phase with a formulation that is not marketable (e.g., not amenable to large-scale production and/or not meeting marketing and commercial requirements) would require the introduction of some formulation changes (e.g., additional excipient and manufacturing processes) later in development. These changes may alter the pharmacokinetics of the initial FFP formulation, therefore requiring additional *in vivo* biocomparison (BC) studies to bridge the FFP formulation with the final marketed product. This could possibly result in additional development costs and in loss of time, impacting the patient life.

It is also worth considering that not all drugs are suitable candidates for FFP formulations. For example, the physicochemical characteristics of the API (i.e., solubility, stability) may prove challenging and hence require a more developed formulation with additional processing steps (i.e., hot melt extrusion (HME)) or excipients (i.e., solubilizer, preservative). Failing to recognize the development risks associated with particular compound could impact the success of the clinical trial and cause further delays before entering in the later stages of clinical development. It is therefore of paramount importance to assess the physiochemical characteristics of the compound before considering the use of FFP strategy.

In the commercial approach, the formulation attributes are similar to those of the final commercial product. The need to develop a marketable formulation and process early on in development is generally driven by the desire of fulfilling the marketing and commercial requirements. However, this approach requires a substantial front-loading of resources, which might be subtracted from other programs. Generally, this might cause a delay in entering clinical trials, as developing a marketable formulation is time consuming. In addition, to support the development activities, a large amount of API would be needed to represent the final scaled-up lot, which could cause additional delays due to timelines associated with API synthesis development. Nevertheless, this approach potentially offers the fastest route to market access as it concentrates the development effort onto a single formulation and reduces, or eliminates, the requirements for BC studies between the initial and marketable formulations. It should also be considered that validation and launch activities are anticipated through this approach, further decreasing the time to market access.

This approach is generally selected for those programs where low attrition is expected and risk of failing the early clinical trial is low. This is the case for those compounds that do not represent a development challenge (e.g., optimal bioavailability, stability) and for which the efficacy and pharmacodynamics are well established (e.g., second generation molecule in an established therapeutic class). For example, the commercial approach might be well suited to a product life cycle enhancement program, where an existing marketed product is reformulated to develop an extended release formulation. Despite the high risk of attrition, resources can also be front loaded to support a more extensive formulation development of challenging compounds (e.g., low solubility, unprecedented pharmacodynamics), which represent potential blockbusters.

Both approaches have advantages and disadvantages (summarized in Table 7.5) that might influence the decision of pursuing one or the other paradigm. These need to be carefully weighted, taking into consideration both the scientific

Table 7.5 Comparison of the merits and demerits of fit-for-purpose and commercial approaches.

Attributes	Fit-for-purpose approach	Commercial approach
Resources needed	Minimal	Large, requires front-loading
API needed	Hundreds of grams	Kilograms
Number of excipients used	Zero to few	Many
Formulation	Simple, tailored for the need of early phase clinical objectives and proof-of-concept studies	Complex, designed to answer marketing and commercial needs
Formulation changes	Expected, might require bioequivalence studies or biowaiver applications	Minimized, similar to final market formulation
Manufacturing process optimization	Not required	Required
Manufacturing scale-up	Not required	Required
Drug product image development	Not required	Required
Dose flexibility	Large	Narrow
Storage	Special storage condition are acceptable (e.g., 2–8°C)	Ambient storage preferred
Stability evaluation	Hours–months	Years
Initial clinical evaluation	Fast tracked	Delayed
Market access	Delayed	Fast tracked

knowledge of the NCE and the business needs before deciding which approach to take. More often than not, the FFP approach is chosen, and recently, specialized equipment has become available to support the manufacture of these simple formulations. Deciding which of the FFP formulations should be selected depends on various factors: availability and stability of the API, dose range, number of units required to support the clinical study, physicochemical characteristics of the drug, and strategic considerations. In the following sections, these factors and their impact on the formulation selection will be discussed, including a review of the various formulation approaches focusing on their merits and demerits.

7.3.2 On-Site Manufactured Clinical Trial Materials

Broadly speaking, the DP used to support an FIH clinical trial can be supplied by using two alternative strategies—the choice of which will depend on a number of factors, such as the trial design (e.g., SAD or MAD), the API's physicochemical characteristics, batch size, and cost of manufacture. In the conventional manufacturing strategy, the finished DP is manufactured, packaged, and released by the sponsor (*in-house*) prior to being shipped to the clinical site where it will be dosed. Alternatively, the manufacture can be considered an extension of clinical activities, and a customized DP (i.e., drug or combination of drugs and excipients) may be extemporaneously prepared and released at the clinical site prior to dosing in accordance with a clinical protocol and manufacturing instructions from the sponsor (i.e., composition, preparation procedure, in-use dating, and storage conditions). Formulations prepared in the latter method can be described as extemporaneous formulations or on-site formulations (OSF).

Provided that formulation development is on the "critical path," using an OSF can accelerate entry into FIH clinical trials by reducing cycle times, resource investment, and expenses. The OSF strategy offers several advantages compared to conventional manufacturing, including:

- Reduces API needs as DP development is minimized due to the reduced stabilization requirements
- Requires only short-term stability data because the formulation is dosed soon after preparation
- Provides the flexibility to adjust the dose, accordingly to the study protocol, to enable response to emerging clinical data
- Expedites the cycle time between quality control and release

The clinical sites do not generally have access to specialized equipment, and therefore an OSF strategy may not be suitable for potent and hazardous drugs or when complex DPs (i.e., tablets, spray-dry intermediates) or large batch sizes are required.

There are differences between the EU and United States regarding the regulatory framework around OSF strategy that the sponsor should consider when planning an FIH study. OSF manufacturing requires adherence to good manufacturing procedures (GMP) guidelines for clinical studies conducted within Europe. The EU clinical directives (Directive 2001/20/EC and Annex 13 of the Eudralex—Volume 4) state that every IMP must be manufactured according to GMP standards by a manufacturing authorization holder. Hence, the manufacture and quality control should be conducted in GMP-compliant facilities in accordance with GMP procedures, and the DP should be released by a QP.

This regulated environment limits the number of sites available and increases the price for conducting an OSF study in the EU compared with the United States, where the DP used in certain Phase I clinical trials are exempt from the requirement of the GMP regulations stated in 21 CFR 210 and 211 (FDA, 2008). The FDA accepts other manufacturing and release assurances, as detailed in the United States Pharmacopeia (USP) <795> (Compounding Non-sterile Preparations), <797> (Compounding Sterile Preparation), <1075> (Quality Assurance in Compounding), and <1191> (Stability Consideration in Dispensing Practices), in lieu of GMP compliance. However, the FDA still exercises oversight on the manufacture of the DPs that are under its IND authority and reserves the right to take appropriate actions if there is evidence of inadequate controls. Under the IND regulations, an OSF strategy is possible for API that is produced according to GMP standards, and the exemption does not apply to drugs that the sponsor has already made available beyond Phase I.

The OSF strategy is generally used for SAD studies and other probing studies, such as taste evaluation, as only a small number of dosing units will be required. Using an OSF may be impractical in MAD studies as the increased demand of manufacture will often be outside of the capabilities of the trial site, as well as incurring a greater expense to the sponsor (Hariharan *et al.*, 2003).

Examples of formulations generally used with an OSF strategy include API in bottle and API in capsule (Hariharan *et al.*, 2003). For API-in-bottle formulations, typically the API is dispensed directly into a bottle and incorporated into a flavored suspending vehicle (such as Ora-Sweet®/Ora-Blend®) on-site to produce a suspension or solution. Solvents or co-solvents can also be added to solubilize the API, if required. This technique does not require extensive formulation development, thereby reducing time and cost investment early on. API-in-capsule formulations can be manufactured on-site by providing the trial site with empty capsule shells and API that requires manual filling into the capsules. However, dispensing particularly low amounts can provide challenges around accurate dose dispensing and is a particularly laborious task. Automated microdose dispensing devices, such as Xcelodose® S, have been developed to aid in the precise dosing of capsules at an increased speed (Capsugel, 2008). Although, this will often be deemed unsuitable in large-scale scenarios further downstream.

In larger-scale MAD studies, capsules and tablets are often more suitable. For this type of formulation, conventional manufacturing processes can be used (e.g., blending, roller compaction, and compression), which provides a greater capability for scale-up compared with OSFs. Favorably, the use of tablets and capsules can often be extended to Phase IIa proof-of-concept studies and can often act as an acceptable starting point for commercial formulation development. This can decrease transition times between trial phases and potentially avoid bridging pharmacokinetic studies downstream, overall creating a more efficient process. However, the *in-house* manufacturing approach can increase the time to dosing in humans and increase cycle times if formulation development is on the "critical path"; this would delay the collection of data and might elongate the early phases of clinical trials.

7.3.3 Solubility

The number of drug compounds with poor aqueous solubility developed has increased, which poses significant formulation design challenges to scientists. Drug solubility and permeability are fundamental API properties that determine the rate and extent of absorption. Knowledge of these characteristics can aid the choice of a suitable formulation.

To characterize and correlate a drug's *in vitro* properties to its *in vivo* performance, a number of models have been devised to aid dosage form design. The Biopharmaceutics Classification System (BCS) (Amidon *et al.*, 1995) has been widely adopted and groups DS into four classes based on its drug permeability and aqueous solubility (Figure 7.7).

Another model, termed the dose number (D_o) concept, that was developed by Amidon *et al.* (1993) calculates the number of intestinal volumes in a biorelevant

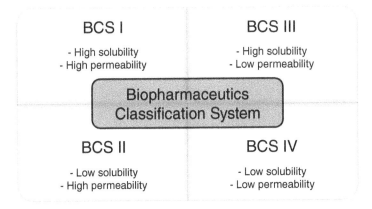

Figure 7.7 Biopharmaceutics Classification System (BCS). Source: Amidon *et al.* (1995). https://deepblue.lib.umich.edu/handle/2027.42/41443.

media (e.g., simulated gastric fluid (SGF) or fasted-state simulated intestinal fluid (FaSSIF)) it would take to dissolve a given dose of drug:

$$D_o = \frac{\text{Dose}(\text{mg})/250\text{mL}}{\text{Solubility}(\text{mg}/\text{mL})}$$

As the value of D_o increases, the relative ability for drug dissolution is reduced, which makes the target exposure increasingly difficult to reach. This is an excellent model for understanding absorption risk and can be used to assess the need for a solubility enhancing strategy. Initially, chemically modifying the drug to improve drug solubility should be explored by the pharmaceutical development team, such as by producing the API in a salt or prodrug form. Effectively designing the API to enhance intrinsic solubility will reduce the need for enabling formulation strategies downstream.

However, if chemical modifications do not improve drug solubility as desired, the choice of a formulation can enhance its dissolution; either a conventional or enabled strategy can be chosen.

7.3.3.1 Conventional Formulations

Conventional formulations use standard manufacturing processes and should not significantly affect drug bioavailability. OSF and *in-house* manufactured DPs can both use a conventional formulation strategy if the API's solubility is deemed acceptable.

For those drugs with inherently good solubility (such as BCS Class I/III or low D_o number compounds), incorporation into an OSF formulation would be an effective strategy as it can be assumed that the dissolution profile of the API would not be impacted by the formulation. If dosage form development is on the "critical path," the use of an OSF will reduce the time to the start of the clinical study and enable the collection of data more quickly. Lower solubility drugs (e.g., BCS Class II/IV or high D_o number compounds) might require functional excipients to aid dissolution and might therefore be more suited to a more developed formulation that is manufactured *in-house* (e.g., spray-dried intermediate). However, if the active drug has poor solubility but an OSF formulation is preferred, a number of enabling strategies to improve dissolution can be adopted.

7.3.3.2 Enabled Formulations

If a drug has inherently low aqueous solubility (e.g., BCS Class II/IV or high D_o value compounds), the use of an enabled formulation to improve aqueous solubility and bioavailability can be considered. Enabled formulations are not often needed for higher solubility drugs as, once the drug has been dosed, it can be expected that rapid dissolution will occur.

A number of traditional methods can be used to "enable" these lower solubility drugs, including physical modifications, as well as excipient changes. If using either an OSF or *in-house* formulation, reducing API particle size as an initial manufacturing step is a common way to improve drug exposure, as increasing the effective surface area exposed to aqueous gastrointestinal fluids increases the rate at which the API is solubilized. Traditionally, this has been achieved by milling the API; however this often imparts little control over particle size and crystalline morphology.

More recent enabling technologies include creation of a solid dispersion by spray drying (Dobry *et al.*, 2009), HME (Breitenbach, 2002; Repka, 2009), or undertaking liquid encapsulation and nano-milling (Peltonen and Hirvonen, 2010). Spray drying is a cost-effective, API-sparing method that can produce a uniform, narrow particle size distribution and can be used for small-scale manufacture, making it suitable for FIH trials. Creation of a spray-dried API intermediate as an initial manufacturing step is often used, which involves dissolving API and carrier (e.g., hypromellose acetate succinate) in a common solvent, solution atomization, and drying, followed by fine powder production and collection (Sollohub and Cal, 2010).

HME involves the melting of an API and an appropriate polymer-based excipient at high temperatures, which are pumped through an orifice with a rotating screw. The exudate produces particles of uniform shape and size when milled (Repka, 2009). However, the high temperatures used within this process can cause drug degradation, particularly if the compound is heat sensitive. In contrast to spray drying, HME is often more suited to larger-scale manufacture and might therefore be unsuitable for the small quantities required within FIH trials.

Other enabling strategies that use nontraditional excipients to enhance solubility can be adopted, which are especially suitable for more developed formulations. Examples include the use of surfactant, complexing agents, liposomes, and self-emulsifying drug delivery systems (SMEDDS), among others (Buckley *et al.*, 2013). For example, SMEDDS are composed of a surfactant, cosurfactant, oil, and API, such that, when in contact with an aqueous phase, it forms an oil-in-water microemulsion. These systems increase the solubility of hydrophobic compounds by presenting the drug in a dissolved form with a small droplet size (Charman *et al.*, 1992).

However, if these technologies are used, it is worth bearing in mind that bridging to a new formulation downstream may be more difficult, and often pharmacokinetic studies will be needed. Also, the cost of manufacture might be increased, and therefore the use of simpler, alternative strategies might be more suitable for this early stage of drug development.

7.3.4 Trial Design Considerations

The formulation strategy chosen should be flexible enough to achieve the target dose range without manufacturing an excessive number of potencies.

If a larger quantity of units is administered, this could cause inconvenience to the patient and contribute toward patient dropouts. Furthermore, the greater the number of units dosed, the greater the quantity of excipients will be administered. It has been found that certain excipients can affect a drug's pharmacokinetics; for example, mannitol (Adkin *et al.*, 1995), which is a commonly used diluent, and xylitol (Salminen *et al.*, 1989) have both been found to accelerate small intestinal transit, reducing *in vivo* drug absorption.

An internal literature review on the development of compound intended for oral use found that in roughly 40% of the non-oncology FIH/SAD studies, the compound is dosed using a liquid formulation (i.e., suspension or solution). The remaining 60% uses solid oral forms, of which roughly two-third uses capsules and the remaining uses tablets (Merck Sharp & Dohme Ltd, Unpublished research).

An API in bottle to extemporaneously prepare a suspension or solution at the clinical site (i.e., OSF) offers the most flexibility as dose adjustments can be achieved by increasing or decreasing the amount of API administered, as well as reducing amount of excipients being administered. For example, concentrations of 1 and 10 mg/mL should be sufficient to dose 0.5 mg–1 g within a SAD study. For doses above 1 g, a higher concentration might be needed, for example, 20–25 mg/mL. It is worth considering whether the lowest and highest volumes will be practical to deliver; volumes smaller than 1 mL and larger than 100 mL could provide administration difficulties.

When solid oral dosage forms are used to support SAD studies, typically between one and three potencies are manufactured, which are dosed as a single unit or as a combination to allow for dose adjustment; examples are provided in Table 7.6 (Merck Sharp & Dohme Ltd, Unpublished research).

7.3.4.1 Placebo Controlled and Blinding

If the FIH trial is to be placebo controlled and blinded, the placebo could, but it is not required to, use the same excipients and manufacturing processes to theoretically create an exact replica of the active formulation. To ensure blinding, the placebo must have the same aesthetics (e.g., shape, color, taste) and administration pattern to the active. For example, if Drug X is an orange, round tablet given once daily at 8:00 am, the placebo should replicate this.

However, problems can arise when the API has unique characteristics not often exhibited by the placebo. For example, bitter active ingredients might exhibit a distinct taste, which could unblind the trial. To resolve this, a taste-masking strategy could be applied (e.g., sugar coat or film coat) or the DS could be encapsulated. For oral solutions, taste can be improved by adding sweeteners or flavorings into the formulation or alternatively making the formulation more bitter, such as by adding Bitrex®.

Table 7.6 The dosage strengths, dose ranges, and number of units used within 10 reported single dose (SD) FIH studies between 1995 and 2014 (Merck Sharp & Dohme Ltd, Unpublished research).

Study type	Oral dosage form	Dosage strength	Dose range	Number of dose units needed to reach dose
Parallel SD	Solution and tablets	Solution: 0.5 mg/mL	0.125–0.5 mg	0.25–1 mL
		Tablets: 1, 2, 4, 8, 16, 32, 64, and 128 mg	1–128 mg	1
	Tablets	50 mg	25–800 mg	0.5–16
	Capsules	10 and 50 mg	30–120 mg	3–4
	Tablets	250 and 1000 mg	750–6000 mg	3–6
	Capsules	25 and 100 mg	25–1000 mg	1–10
	Capsules	125 and 250 mg	125–1000 mg	1–4
	Capsules	125 and 250 mg	250–750 mg	1–3
	Capsules	50 and 200 mg	200–1200 mg	4–6
	Tablets	300 mg	2100–4500 mg	7–15
Crossover SD	Capsules	2 and 25 mg	2–25 mg	1
	Capsules	0.1, 1, and 5 mg	0.1–40 mg	1–8

To ensure blinding remains, taste testing of the formulation can be carried out by using a human-taste test panel. However, concerns around the subjective nature of this test are apparent, and within recent years electronic tongue evaluation (Latha and Lakshmi, 2012) and animal studies (Soto *et al.*, 2015) have become available. The electronic tongue works by detecting organic and inorganic molecules and transforming the data into electrical signals, which are interpreted into a taste profile. It can undertake a number of functions, including quantifying the bitterness of actives, developing a placebo with comparable bitterness, and suggesting a suitable taste masking approach, among others (Murray *et al.*, 2004).

7.3.5 Types of Formulations

7.3.5.1 Ready-to-Use Solutions and Suspensions

As previously discussed, these formulations consist of a solution or suspension of the API in a liquid vehicle. The vehicle is generally represented by water or buffer systems and might contain water-miscible solvents (e.g., glycerin, ethanol) to improve the wetting and solubility of the API. When a suspension is formulated, polymers (e.g., carboxymethyl cellulose or methylcellulose) might be added as thickening agents to reduce the sedimentation of the API. In addition, flavors and sweeteners might be added to mask the taste of the API.

If prepared in-house and then shipped to the clinical site, the formulation often contains preservative(s), to improve the microbiological stability. Conversely, preservative(s) is typically not required when a solution or suspension is extemporaneously manufactured at the clinical site (i.e., OSF) from an API in bottle formulation.

The equipment needed to manufacture these formulations is very simple, and the process is neither complicated nor time consuming. Simply, an overhead stirrer or high shear mixer could be used for preparing a solution or a suspension, respectively. The formulation could then be dispensed into bottles using either a peristaltic pump or calibrated pipettes. Clear bottles can be used to ease the evaluation of appearance; however, amber bottles are generally the standard to protect the API from light degradation.

Assuming that there are no safety concerns for its use in humans, the first choice would be to adopt the formulation used during the toxicological preclinical studies. This would allow leveraging of the preclinical stability data, hence considerably cutting development times and providing a faster way to enter the clinical study.

Ready-to-use suspensions and solutions have the additional merit of providing high dose flexibility and are therefore desirable during SAD studies. The formulation in fact could be diluted, using the same vehicles, and dispensed by the pharmacist or the clinician at the clinical site just before dosing. An additional advantage of using ready-to-use suspensions and solutions is the possibility to treat pediatric and geriatric populations or those patients that have swallowing difficulties. The main disadvantage of these formulations is that they do not represent a valid alternative going forward in development and do not represent the preferred formulation as commercial products, unless the product is targeted to population with difficulties in swallowing.

When manufactured in-house, microbiological stability data should be generated to ensure that microbial growth remains within an acceptable level for the duration of the clinical study. Dosing accuracy should be also evaluated, and rinsing tests should be performed to demonstrate that the right dose is delivered at time of dosing. The taste of the API might be also a challenge as this is more perceptible for liquid formulations. Taste evaluation studies might be needed to mask the bitterness of the API and to develop a matching placebo. These studies could be run in conjunction with the pharmacokinetic studies or alternatively with electronic tongue evaluation (Latha and Lakshmi, 2012) and animal studies (Soto *et al.*, 2015).

7.3.5.2 API in Bottle

Ready-to-use suspensions and solutions manufactured in-house are often not amenable to support clinical trials due to the short stability of the API in the liquid vehicles (e.g., sedimentation, degradation). A simple alternative to this issue is to use API in bottle formulations. Ideally, this approach relies on gravimetrically dispensing the neat API powder directly into vials or bottles.

The API is then extemporaneously reconstituted into the liquid formulation at the clinical site using an appropriate vehicle. This could be represented by the vehicles used for the preparation of the preclinical formulation or alternatively by marketed products such as Ora-Sweet and Ora-Plus® (Fish *et al.*, 1999), as well as with fruit juices (Sistla *et al.*, 2004).

As previously mentioned, API in bottle formulations is generally used in conjunction with an OSF strategy. In this case, the neat API is bulk dispensed into a small number of bottles and shipped to the clinical site. At the clinical site, a portion of the API is then dispensed from the bottle to prepare the extemporaneous formulation to be dosed.

When an API in bottle formulation is not used in conjunction with OSF strategy, a large number of bottles are needed to be filled prior to shipping to the clinical site. The containers can be filled manually but this option can be laborious and time consuming. Alternatively, equipment is available to semiautomatically or automatically dispense the API into vials or bottles. This equipment gravimetrically dispenses the powder into a container (i.e., bottles or vials) that is placed onto a microbalance. Using automated equipment, weights between 100 µg and 300 mg, with a relative standard deviation (RSD) of less than 2%, can be dispensed at a speed of hundreds of units per hour. The most widely utilized of such equipment is the Xcelodose (Capsugel, Morristown, NJ, USA), but also other less known alternatives are available such as the FillPro Net-Weight® (FillPro, Inc., Golden, CO, USA), Powdernium® (Symyx Technologies, Santa Clara, CA, USA), and Quantos® (Mettler-Toledo, Columbus, OH, USA) (Tablets & Capsules, 2009).

The functional core of the Xcelodose system is composed of a dosing head and a microbalance. The dosing head dispenses the powder directly into the bottle—but also vials and capsules can be used—through a tapping mechanism to aid the flow of the powder. The microbalance serves two functions: (i) it provides the accurate control of the action of the dosing head, by stopping its function when the target weight is reached, and (ii) it checks the weight of each filled bottle. The weight of each filled bottle is recorded, and those bottles that are outside the acceptance criteria are automatically rejected by the Xcelodose. The weight checking/sorting of the bottle is hence integrated with the dispensing operation, eliminating the need for post-processing operations.

Either neat API or powder blend can be dispensed. The first route has the major advantage of eliminating the need for formulation development activities, excipient compatibility, and content uniformity studies, allowing a faster production and release of the clinical supply. The method development for the stability studies is also straightforward, as neat API is used, and interferences from other components are eliminated. In addition, this approach has the merit of minimizing the amount of API required, which could be advantageous when API is scarce. However, for this approach to be successful, the API should be non-cohesive and free flowing—unless manual weighing is used. Alternatively, utilization of physical processing steps (e.g., spray drying or granulation) or addition of flow-aiding excipients (e.g., glidants) could be used to improve API characteristics.

Excipients might also be needed to improve the solubility (e.g., solubilizers or wetting agents) if reconstituted as solution or resuspendability (e.g., wetting agents) if reconstituted as suspension. Alternatively, the formulators could investigate the effect of the particle size on the suspendability and dissolution of the API (Sistla *et al.*, 2004). Additional excipients may also be necessary to improve the bioavailability and stability of the API. The taste of the API could also represent a challenge, and taste evaluation studies might be needed to mask the bitterness of the API and to develop a matching placebo.

Using the API in bottle approach permits high dose flexibility because this would require only changing the amount of powder that has been dispensed. However, care should be taken to assess whether changing the amount of powder negatively impacts the reconstitution step as varying the ratio between API and vehicles might impact solubilization and dispersion of the active.

The stability of the API (or powder blend) in the bottle (i.e., powder prior to reconstitution) should be assessed; if neat API is used, it is acceptable to leverage the stability data of the DS for the clinical trial application. The stability of the extemporaneously prepared suspension or solution should be also assessed. A short-time stability study would be sufficient, as long as it allows enough time to cover the preparation and the dosing of the extemporaneously prepared formulation. Reconstitution studies should be performed to make sure that the powder can be easily solubilized, or suspended, at the clinical site. In addition, rinsing studies should be performed to ascertain the dosing accuracy. The effect of the reconstituting vehicles on the bioavailability of the API should not be overlooked as well; for example, the vehicle could impact the dissolution rate of the API due to its viscosity (Aubry *et al.*, 2000) or inhibit metabolic enzymes as seen with grapefruit juice (Becquemont *et al.*, 2001).

When using the API in bottle approach, the time to develop and produce the clinical supply can be reduced to less than 2 months (Sistla *et al.*, 2004). However, the approach has the downside of devolving the responsibility of the final formulation preparation to the clinic. This not only increases the resources needed at the clinical site but also requires the personnel at the clinical site to be adequately trained to prepare and dose the formulation within an acceptable time after preparation. Clear instructions should be provided to the clinical site on how to prepare the final formulation and on how to recognize possible problems.

7.3.5.3 API in Capsule

An analogy to the powder in bottle approach is to directly dispense the neat API into capsules. Both OSF and *in-house* strategy can be used to manufacture API in capsule formulations. The manufacturing process is similar to that used for powder in bottle, with the difference that API is directly dispensed into a capsule shell rather than a bottle or vial. Historically, the API was gravimetrically dispensed into capsules via manual operation, but this method is laborious and time intensive. Alternatively, capsules can be filled using automated equipment such as the Xcelodose system, which is able to fill hard gelatin or HPMC capsules from size 00 to 4 with great accuracy (i.e., RSD <2%) and minimal waste of API. Similar to powder in bottle manufacturing, the Xcelodose system eliminates the need for a separate weight-sorting step as the weight of each capsule is checked, and those capsules that are outside the acceptance criteria are rejected.

Dispensing API directly into capsules allows high dose flexibility as simply varying the amount of API dispensed into the capsule can modify the dose. The development timeline can be greatly reduced because eliminating the use of excipients reduces the resources needed for compatibility studies, formulation development work, and content uniformity testing. Notably, this also reduces the amount of API needed for the development of the clinical supply.

Analytical development work can be straightforward because a simplified assay and impurity method can be used due to the absence of additional ingredients. In addition, there is no requirement to assess the stability of the DP. Since the capsule contains only the API, it is considered acceptable to leverage the stability data of DS for the clinical trial application.

Dispensing the API into capsules offers a solution to some of the challenges associated with the use of powder in bottle. The API is contained within the capsule shell, eliminating the need to perform taste-masking studies. Blinding of these studies is straightforward as well as the development of a matching placebo. Notably, identically looking capsules can also be used to over-encapsulate comparator products to support "active control" trials. Since the capsule is dosed as it is, there is no need to develop a suitable vehicle, consequently eliminating the need for performing stability studies on reconstituted product and rinsing studies.

It should be noted, however, that dispensing neat API into capsules through automated equipment might not be feasible for every compound. The API should have adequate flow characteristics (i.e., free flowing and non-cohesive), be stable enough once encapsulated, not interact with the capsule shell (e.g., moisture sensitive, hygroscopic, or containing aldehyde groups), and have adequate solubility. The API in capsule approach is generally preferred for BCS Class I and III molecules; however, when used with BCS Class III, the addition of absorption enhancers (e.g., surfactants) may be considered to increase the bioavailability of the drug. An API in capsule approach is not indicated for those compounds that have poor solubility in physiological fluids and therefore requires excipients to increase their solubility. Before selecting this approach, it is important to assess the biopharmaceutical characteristics of the compound to avoid negatively impacting the quality of the clinical data due to suboptimal exposure of the compound. An alternative that could be used for low solubility compounds (BCS Classes II and IV) is to formulate and dispense a spray-dried intermediate of the drug to increase its solubility.

In addition, it should be considered that the Xcelodose system can only dispense weights between 100 µg and 300 mg; therefore this approach is not indicated for potent drugs with dose less than 100 µg. This challenge could be simply overcome using a powder blend to dilute the API. However, segregation of the powder, and consequently content uniformity issues, might occur due to the tapping action of the dosing head. It has been shown that segregation could be minimized by formulating a cohesive powder blend through physical processing (i.e., co-micronization or granulation) or through the selection of appropriate diluent (i.e., particle size similar to that of the API) (Bi *et al.*, 2010).

The challenges previously mentioned might be overcome by adding additional excipients or pre-encapsulation steps (granulation, milling, or spray drying). However, this would require extra development activities, which would retard the start of the clinical study, hindering the advantages of using an API in capsule approach. When these challenges arise, the project might be diverted toward a more conventional formulation, such as formulated tablets or capsules, which represent a prototype of the final marketable formulation.

7.3.5.4 Formulated Capsules and Tablets

The use of more conventional formulations to supply FIH studies is more resource intensive compared with the other approaches discussed so far. Formulating a dry-filled capsule (DFC) or tablet requires the use of various excipients (e.g., diluent, disintegrant, and lubricant) and a wide range of formulation activities. Also, the analytical effort to characterize the dosage form is increased because additional tests are required, such as disintegration and dissolution.

It should be considered that, as a result of this wide range of activities, a larger amount of API is required compared with the previous approaches. This amount is further increased when additional strengths of the formulation are required for the dose escalation studies. Before considering this approach, it is advisable to determine up front the amount of API that is required. It is also important to work together with the chemical development team to avoid scarcity of the API that would delay the beginning of the formulation activities or impact their progression. It should be also considered that the variation to the API critical attributes should be minimized to avoid impacting the performance of the developed DP. Therefore, it is generally preferable to delay the start of the formulation development until there is confidence that the API critical attributes will not drastically change due to the scale of the synthetic process.

This front-loading of resources can be balanced by a faster progression through the subsequent phases of clinical evaluation. It is in fact easier to supply larger clinical trials with formulations that can be manufactured using pilot-to-commercial scale equipment. In addition, this approach may eliminate the need for BC studies from the formulation used for the FIH study, as long as the critical attributes of the DS and product remain the same throughout the clinical development.

The development of capsules is generally more straightforward than tablets because compressibility and taste of the API are not an issue. A strategy to expedite the development of immediate release products is to prepare simple blends of API with limited number of excipients (i.e., API, diluent, disintegrant, and lubricant). The blend selected should serve the purpose of diluting the API and provide sufficient flowability to be amenable for volumetric encapsulation while providing adequate exposure when dosed. It should be noted that when developing a range of formulations with different strengths, various blends would need to be developed. These blends can then be volumetrically encapsulated in hard gelatin, or HPMC, capsules using manual equipment such as the FASTLOCK filler (Feton International, Brussels, Belgium), MF-30 (ACG North America LLC, South Plainfield, NJ, USA), and KAP-300 (Dott. Bonapace & C, Limbiate, Italy), which are capable of producing a few thousand capsules per day for use in the early clinical studies.

When using these simple blends, analytical development can also be straightforward; only a limited number of excipients are present, and this limits the interferences on the chromatographic method. This approach, on one hand, reduces the resources required for formulation and analytical development, while, on the other, generating some useful preliminary data on excipient and capsule compatibility that could be used as a starting point for the commercial formulation.

DFCs are generally filled volumetrically, and, therefore, the internal volume of the capsule shell limits the amount of API that could be encapsulated; this can represent a challenge when the dose required is high. In this case, granulating the blend, through roller compaction or wet granulation, is generally an effective way to increase the bulk density of the powder to be encapsulated. Granulation also has the additional advantages of improving powder flow and content uniformity, but it requires additional resources for manufacturing and development. Alternative to granulation, the blend could be directly compressed to form tablets.

The limited internal volume of the capsules can also represent a challenge when large quantities of excipients are required and the drug loading of the blend is low. This is often the case when HME or spray-dried intermediates are used to improve the bioavailability of the compound. In these situations, developing a tablet formulation is generally preferred. When developing a tablet, the compressibility and flowability of the powder, as well as the mechanical properties of the finished product, are critical for success. If producing a variety of potencies, the formulator could also consider using a common blend, while varying the tablet image, to support the production of the various doses. This could save the time to develop the additional blends but would require additional resources to manufacture a matching placebo. Finally, the taste and color of the API might also represent a challenge when developing tablets, and a coating step might be needed to facilitate the development of a matching placebo. This additional process can increase the development time, further delaying the supply of the clinical batch.

Table 7.7 summarizes the main advantages and disadvantages of the various formulations available for use in FIH studies that have been described in the

Table 7.7 Pros and cons of the various formulations available for use in FIH studies.

Dosage form	Pros	Cons
Ready-to-use solutions and suspensions	Simple manufacturing process and equipment needed	Not amenable for large-scale production
	Formulation and analytical development might be straightforward if the preclinical formulation is used	Chemical stability of the API might be an issue
	High dose flexibility	Microbiological stability might be an issue
	Possible to treat patients that have difficulties in swallowing	Preservatives might be required
		Dosing accuracy (i.e., rinsing study) should be evaluated
		Taste and smell of the API might be a challenge
API in bottle	Fast production and release of the clinical supply	Excipients may be needed to improve the solubility and resuspendability
	Could be extemporaneously prepared at clinical site	Dosing accuracy (i.e., rinsing study) should be evaluated
	High dose flexibility	Taste and smell of the API might be a challenge
	Automated equipment available	Reconstitution vehicle should be developed or selected
	Possible to treat patients that have difficulties in swallowing	Physician at clinical site needs to be trained
	Only short-time stability study in liquid media required	
API in capsule	Taste is not a problem	Challenging for large-scale production (e.g., Phase II onward)
	Easy to develop a matching placebo; also active control is possible	Not amenable for API with low solubility
	Automated equipment available	The API should be free flowing
	If excipient are not used, the stability data of drug substance could be used for the filing	Not amenable for potent drug

	Advantages	Disadvantages
Dry-filled capsules (DFCs)	Taste is not a problem	Lower dosing flexibility compared with API in bottle and ready-to-use solutions/suspensions
	Easy to develop a matching placebo; also active control is possible	Delayed start of the clinical study
	Automated and semiautomated manufacturing process available	Larger amount of API is required
	Capsules are generally more straightforward to develop than tablets	Reduced dosing flexibilities
		Challenging when drug loading of the blend is low
Formulated tablets	Acceptable starting point for commercial formulation development	Delayed the start of the clinical study
	Automated manufacturing process	Large amount of API required
	Greater capability for scale-up, faster progression to subsequent clinical phases	Reduced dosing flexibilities
	Reduced risk of having to perform biocomparison studies	Challenging when drug loading of the blend is low
		Resource-intensive development process
		Taste and color of the API might be a challenge

previous sections. The selection of the formulation to initiate the FIH study should be based on strategic factors as well as on the basis of a careful evaluation of the merits and demerits of each formulation in consideration of the characteristic of the molecule to be formulated.

7.3.6 Formulation Development Strategy

A common strategy employed during formulation development of an FFP formulation is to develop a prototype, then probe, and finally clinical batches of the formulation. This stepwise approach enables the formulator to critically assess the viability of the formulation at smaller scales before committing to a larger clinical manufacture.

The aim of the prototype batch is to determine if the designed formulation has got appropriate content uniformity and dissolution characteristics. The content uniformity is a fundamental quality attribute that is used to demonstrate a constant dose of drug between individual capsules or tablets. Dissolution is generally not assessed as a quality attribute at this stage of development due to a lack of any *in vivo* clinical data to build an *in vitro–in vivo* understanding. However, it can be useful to assess the dissolution properties of the formulation(s) to aid decision making during formulation development. For example, biorelevant dissolution tests can be useful to assess API particle size effects and potential need for surfactant inclusion when formulating BCS Class II molecules (Amidon *et al.*, 1995). Prototype batches are typically manufactured at a small scale, for example, a few grams.

When the formulator is satisfied with the analytical results for the prototype formulation, then he can move on to manufacturing a probe batch (batch scale ~100–200 g). The primary purpose of the probe batch is to generate 4 weeks of stability data at accelerated conditions to determine if the formulation is viable from a stability perspective. Finally, if all the analytical data looks good at this stage, then it is appropriate to proceed with the larger clinical manufacturing campaign.

If project timelines are very tight, then it is feasible to compress development timelines by either combining the prototype and probe formulations or manufacturing the clinical batch "at risk" (i.e., without knowing the formulation stability profile). Information from the excipient compatibility studies and from knowledge of the DS stability profile should enable a reasonable assessment of the risks associated with manufacturing the clinical batch without knowing the formulation stability profile.

Where minor stability issues are observed for the formulation, a 2–8°C storage strategy can be employed to help mitigate the issues. The drawback of this approach is the cost and extra resource required to transport supplies under refrigerated conditions to the clinic. The DP will also need to be handled at 2–8°C at the clinic, which can make management of dosing schedules more difficult.

7.3.7 Bridging Studies to Commercializable Formulation

During the clinical development of a new compound, it is common for the API and the formulation attributes to undergo various changes. Such changes are generally the result of an optimization process aimed to improve the manufacturability and the clinical performance of the DP. As we have seen in the previous sections, these changes are unavoidable when developing an FFP formulation to support the initial phases of clinical evaluation.

When designing an FFP formulation, it is important to consider the ease of transitioning to a new formulation in later clinical trial phases. A line of sight to the commercial dosage form should be considered; for example, if the drug is intended to be combined with another drug(s) or drug regimen downstream, then using the same (or similar) excipients to the established medicine(s) could be implemented. This would aid the transition of bridging to a fixed dose combination later down the line, which could be especially important if targeting those conditions that require combination therapy, for example, HIV or tuberculosis.

Changing the critical attributes of the API (e.g., salt form and particles size), or of the dosage form (e.g., excipient used and formulation type), might impact the pharmacokinetic properties of the formulation and consequently its safety and efficacy. To minimize the risk of carrying forward a suboptimal formulation to the subsequent clinical study, BC or BE studies are generally performed, which aim to assess the impact that a change of formulation has on the drug's biopharmaceutical performance. The main aim is to demonstrate that any changes have not impacted the bioavailability of the drug. These supplementary trials can be used to aid decisions around dose selection in the new formulation for Phase II trials, as well as advising the development of a marketable formulation downstream. In BE/BC studies, the systemic exposure profile of a reference DP is compared with that of the new developed formulation (FDA, 2003). BC studies are generally used prior to a pivotal study (e.g., Phase IIB for Pharmaceuticals and Medical Devices Agency (PMDA) or Phase III for FDA and EMA), when there is no regulatory requirement to show comparability between the formulations; hence, they are strictly used to mitigate a business risk. Conversely, for any formulation changes implemented after the pivotal study, it is a regulatory requirement to demonstrate that the changes have not impacted the systemic exposure profile of the formulation (FDA, 2003). However, it is important to note that, besides being a regulatory requirement, BE studies are also a useful tool for pharmaceutical companies to minimize the business risk of progressing forward a nonoptimal formulation.

When undertaking a BC/BE study, a lower dose should be administered rather than the previously determined safe dose for humans. A sufficient safety margin is advised to account for (and prevent any) unexpected safety concerns due to enhanced drug exposure. For example, a published Phase I BC study evaluated drug pharmacokinetics and exposure in an OSF oral suspension compared with a DFC. They dosed at a potency of $81 \, \text{mg/m}^2$ even though the predetermined MTD was $120 \, \text{mg/m}^2$ (Chatterjee *et al.*, 2005).

BE studies can be expensive (i.e., up to $250 000) and time consuming (i.e., up to 2 months for completion); therefore, their impact on the timeline and resources allocated to the clinical development of a drug must be taken into consideration. When developing an immediate release oral formulation, in order to save the time and cost needed to perform BE studies, pharmaceutical companies can request a "biowaiver" from a regulatory agency. A biowaiver is an exemption (i.e., waiver) from conducting the *in vivo* BE studies, which is granted when *in vitro* tests could be leveraged as a surrogate to demonstrate that the two DPs are therapeutically equivalent (EMA, 2010; FDA, 2000).

The BCS of an API is used as a scientific basis to request a biowaiver (Benet, 2013; Yu *et al.*, 2002). For BCS Class I and III compounds, it can generally be assumed that the change in formulation composition would not significantly impair drug absorption *in vivo*. This can support a biowaiver to avoid demonstration of pharmacokinetics. Specifically, a drug needs to be categorized as BCS Class I or III to qualify for an EMA biowaiver. Notably, the FDA requires the drug to be categorized only as BCS Class I to qualify for a biowaiver; however, BCS III drugs are currently taken into consideration in an FDA draft guidance (FDA, 2015). Conversely, for BCS Class II and IV compounds, where dissolution is the rate-limiting step and obtaining sufficient drug exposure is a greater issue, the impact of formulation changes on pharmacokinetics is less easily predicted. A BE study is often needed in this situation to determine a dose for the new formulation, which ensures adequate bioavailability without safety concerns.

In addition to the BCS, to classify for a biowaiver, it should be demonstrated that both the reference and test formulations show comparable (i.e., similarity factor $f2 \geq 50$) dissolution profiles in three different media: (i) buffer pH 1.2—SGF without enzyme or 0.1 N HCl, (ii) buffer pH 4.5, and (iii) buffer pH 6.8 or simulated intestinal fluid without enzyme. Both products should also demonstrate rapid (85% within 30 min) or very rapid (85% within 15 min) dissolution. Notably, when BCS Class I drugs show very rapid or rapid dissolution in all three media, a profile comparison is not required. Only very rapidly dissolving DPs containing BCS Class III drug are eligible for a biowaiver.

It is also required that the DPs only contain excipients that have been previously used in other approved immediate release formulations and that none of the excipients could affect the rate and the extent of absorption of the active compound. Drugs with a narrow therapeutic index do not qualify for a biowaiver, as well as drugs that are preferentially absorbed in the oral cavity.

7.4 Analytical Development

The analytical sciences department is responsible for the analysis of DP manufactured during the FFP development process. The scope of the department includes identifying the testing requirements, defining test specifications, early development and validation of analytical methods, defining a stability testing schedule, undertaking stability testing, microbiological evaluations, establishing the product reevaluation date (RED), cleaning verification testing, and release testing of the clinical product.

7.4.1 Overview of Analytical Activities during Early Phase Formulation Development

In the formulation section, a common approach to early phase formulation development of prototype, probe, and clinical manufacture was defined. The formulation section also briefly mentioned some of the characterization tests performed at each stage of the formulation development paradigm. These analytical activities are discussed in more detail in Table 7.8.

Table 7.8 Overview of analytical activities typically performed during development and clinical manufacture of FFP formulations.

Formulation type	Analytical activity	Description
Prototype	Content uniformity testing	Test used to demonstrate a constant dose of drug between individual dosage units
	Dissolution (may include testing using biorelevant conditions)	Used to assess amount of API being released from the dosage form over a period of time. Provides insight into formulation performance differences and can be helpful for directing formulation development activities
Probe	Method development and validation	Drug product (DP) analytical methods that are required for generating data for the CTA/IND filing should be developed and validated to the appropriate level defined by the regulatory authority
	Initial batch analysis using the DP specification tests	Initial batch analysis data generated on the DP for each of the specification tests (Section 7.4.2) listed in the filing. The probe data can be used for the initial CTA/IND filing if batch analysis data for the clinical batch is not available at the time of filing

(Continued)

Table 7.8 *(Continued)*

Formulation type	Analytical activity	Description
	Stability: batch analysis after X weeks (e.g., 4 weeks) accelerated storage conditions, for example, 40°C/75% RH	Data generated on the DP stored at various storage conditions using the specification tests (Section 7.4.4) listed in the filing. The data is used to help determine if the formulation is viable to move forward for clinical manufacture and to set initial reevaluation date and storage conditions. The data can also be used in the CTA/IND filing as long as studies on the clinical batch are also commenced prior to initiation of the clinical study
Clinical	Method development and validation	This work should be performed prior to analysis of the clinical batch if the analytical methods have not been suitably developed and validated for the probe batches
	Initial batch analysis and analytical release testing	Initial batch analysis data generated on the clinical DP for each of the specification tests (Section 7.4.2) listed in the filing. This data should be added to 3.2.P.5 of the filing (Section 7.5.4). A certificate of analysis (CoA) should also be generated summarizing the data and compliance to the stated specifications. The CoA is used by the quality control organization as part of the batch certification process prior to final release
	Clinical stability studies	Batch analysis of samples stored under accelerated and long-term conditions at various time points. The data is used to determine product shelf life and storage conditions

7.4.2 Analytical Method Development

The analytical method requirements for early phase product development follow the same principles as late phase development and commercialization. However, in early phase development the understanding of the API characteristics, API stability, and product performance will be limited. Project timelines will generally dictate the level of method development time available to the analytical group. Timelines are generally tight; therefore, it is important to leverage any available API solubility data, API-forced degradation studies, and analytical methods previously developed during release of the API.

The analytical method development process will not be discussed in depth here; however, the API analytical methods will typically be a good starting point for early phase products. Further method development may also be needed due to potential interferences from excipients and new impurities or degradation products caused by the presence of excipients.

Method development activities could include minor chromatographic changes to full chromatographic method screens, establishment of sample extraction solvent and parameters, and filter validation studies. The analytical methods are used to ensure product stability over the time the supplies are required in the clinic, and therefore, assay/degradation methods need to be stability indicating and have adequate sensitivity to detect degradation to below reporting thresholds (limit of quantitation (LOQ) typically <0.5% of API label claim).

The initial methods will require further development through the product life cycle due to a formulation switch to the commercial product, method robustness issues, or general improvements. These changes must be managed in a controlled manner via a change management system. Quality by design (QbD) principles and methodologies are generally not applicable to DP used in FIH study as the formulation type and strategy are likely to change significantly through the product life cycle; hence it is questionable whether investment of resources into QbD at this stage is appropriate.

7.4.3 Analytical Method Validation

Analytical method validation must be performed, for regulatory submitted methods, to demonstrate that the analytical method is suitable for its intended use. Analytical method validation should follow ICH Guideline Q2(R1), and for FIH, Phase I and Phase IIa, the following validation tests are considered (ICH, 2005):

- Accuracy—the closeness of agreement between the value that is accepted either as conventional true value (or an accepted reference value) or as value determined through experimentation
- Linearity—the ability of an analytical procedure within a given range to obtain test results that are directly, or by a well-defined mathematical transformation, proportional to the concentration (amount) of analyte in the sample
- Precision—the closeness of agreement (degree of scatter) between a series of measurements obtained from multiple sampling of the same homogenous sample
- Repeatability (intra-assay precision)—the precision under the same operating conditions over a short interval of time
- Specificity—the ability to assess unequivocally the analyte in the presence of components that may be expected to be present, such as impurities, degradation products, and matrix components
- Quantitation limit (limit of quantitation)—the lowest amount of analyte in a sample that can be determined with an acceptable precision and accuracy under the stated experimental conditions.
- Solution stability—length of time the sample and standard solutions are considered stable for analysis

The tests for quantifying assay, impurity, degradate, and content uniformity values of the formulations and their associated acceptance criteria are summarized in Table 7.9.

Table 7.9 Examples of first-in-human/phase I analytical method validation tests and requirements.

Parameter	Assay/impurities/degradates		Uniformity of dosage units	
	Procedure	Acceptance criteria	Procedure	Acceptance criteria
Accuracy—parent level	No testing is required OR Recovery of active from spiked placebo at 100% in triplicate	Inferred by precision, linearity, and specificity OR Mean recovery: 98–102%	No testing is required OR Recovery of active from spiked placebo at 100% in triplicate	Inferred by precision, linearity, and specificity OR Mean recovery: 97–103%
Accuracy—degradate level	Quantify duplicate preparations of the 0.1% standard solution	Mean is 70–130% of the target value	N/A	N/A
Linearity—parent level	Solutions of the active in duplicate, at a minimum of three levels including 100% (50–120%)	Correlation coefficient >0.990	Solutions of the active in duplicate, at a minimum of three levels including 100% (50–130%)	Correlation coefficient >0.990
Linearity—degradate level	No testing is required	Inferred from the 0.1% quantification from accuracy–degradate level	N/A	N/A
Repeatability—measurement precision	At least five replicate measurements of a standard solution	RSD ≤1%	At least five replicate measurements of a standard solution	RSD ≤1%
Repeatability—method precision	Duplicate sample preparations	Spread between the individual assay active results is ≤5.4%	Test 10 content uniformity samples	RSD ≤6.0%
Specificity	Demonstrate resolution and lack of interference with peaks of interest	Interference with the active peak and critical degradation products must be ≤0.1% or reporting threshold	Demonstrate resolution and lack of interference with the active	Interference with active must be ≤2%
Quantitation limit (LOQ)	Calculate LOQ from the sensitivity standard solution	LOQ ≤ reporting threshold	N/A	N/A
Solution stability	Reassay aged standard and sample solutions against a freshly prepared standard solution	Standard solution: 98.0–102.0% Sample solution: 98.0–102.0% and ≤0.1% (absolute) degradate growth	Reassay standard and sample solutions against a freshly prepared standard solution	Standard solution: 97.0–103.0% Sample solution: 97.0–103.0%

Depending on the formulation type, the planned study requirements and the potential patient population, analytical product testing, and validation of the analytical methods to a higher standard may be deemed appropriate. This will allow the batches formulated and manufactured for FIH/Phase I studies to be used into later phase studies. This may also include analytical methods used for internal requirements that will not be registered at this stage of development (e.g., dissolution/microbial limits tests).

7.4.4 Drug Product Specification

The product specification is a list of tests including associated acceptance criteria and references to the analytical test methods. It establishes the set of criteria to which a new DP should conform to be considered acceptable for its intended use. If a DP meets the listed acceptance criteria, when tested according to the listed validated analytical procedures, it is said to "conform to specifications." Specifications are critical quality standards that are proposed and justified by the manufacturer and approved by regulatory authorities as conditions of approval.

The specification can include regulatory filed acceptance criteria and analytical tests but also any applicable internal acceptance criteria or analytical tests that are performed. Internal tests are generally used to provide additional information to help guide development activities. They can be either more stringent limits that fall within the boundaries of the regulatory filed acceptance criteria or additional tests that are not part of the regulatory specification. Failing to meet the internal specification can be an early indicator for undesired product changes while on stability testing such as degradation, dissolution changes, and form changes. While the product still meets the regulatory filed specifications, these early indications generally trigger further investigation, which can help define future product development activities and improvements.

The specifications for a new DP in the early phase will follow ICH Guideline Q6A (ICH, 1999) and ICH Q3B (R2) (ICH, 2006a). Experience gained in DS or DP development will guide the setting of the DP specification. However, since only limited data is available at this stage of development, the setting of DP specifications for FIH or Phase I studies will be set based on general guidance rather than specific product quality attributes. Specifications will be expected to be revised during the drug development life cycle as new information and data becomes available. The tests generally required for all new DPs are:

- Description (e.g., size, shape, color)
- Identification
- Assay
- Impurities/degradation products

Table 7.10 Summary of analytical testing performed on solid and liquid FIH formulations.

Test	Tablet/ capsule	Solution/suspension/powder for reconstitution
Description	X	X
Identification	X	X
Assay	X	X
Impurities/degradation products	X	X
Uniformity of dosage units	X	X
Disintegration	X	—
Dissolution	I	—
Microbial limits	I	I
Physical form	I	I

X, analytical test must be performed; I, analytical test may be performed, if required, for example, physical form testing for an amorphous formulation.

For more complex formulations such as liquid-filled capsules, amorphous dispersions, and formulations containing salts, additional tests will be required. These may include:

- Uniformity of dosage units
- Disintegration
- Dissolution
- Water content
- Microbial limits
- Polymorphic form

A summary of the tests typically performed on FFP solid oral formulations and OSF liquid formulations (i.e., powder in bottle for reconstitution) is given in Table 7.10.

For FIH or Phase I solid dosage forms (API in capsule, tablets, formulated capsules, etc.), disintegration can be used instead of dissolution as an appropriate test for a regulatory filed specification and test since the dissolution test and data cannot be linked to any *in vivo* clinical data at this point in DP development. With the aim of early formulation development to produce a formulation that breaks up quickly (immediate release formulation), disintegration is deemed an adequate measure of this physical product attribute. Dissolution data are generally collected as an internal non-regulatory filed test to assess against *in vivo* data to be able to set appropriate dissolution specifications later in the product development cycle.

7.4.5 Initial Drug Product Batch Analysis and Release Testing

Following the manufacture of the probe batch, initial batch analysis data is generated using each of the tests listed in the DP specification. This data is then evaluated for compliance with the specification and associated acceptance criteria. If the results meet specification, then they can be used in the "Control of Drug Product—3.2.P.5" section of the regulatory filing (Section 7.5.4).

Release testing is performed after the manufacture of the clinical batch. Release testing is essentially the same as initial batch analysis, except that the data generated is summarized in a Certificate of Analysis (CoA). The CoA is then used by the quality control organization as part of the batch certification process prior to final release of the batch to the clinic. A detailed description of the batch certification process can be found in Annex 13 of the Eudralex (European Commission, 2010a) and 21 CFR 211 (FDA, 2008). DPs failing to meet established standards or specifications and any other relevant quality control criteria will be rejected; however reprocessing may be performed if this is deemed to be appropriate.

7.4.6 Drug Product Stability

Stability data are required in all phases of the IND or CTA to demonstrate that the DS and DP are within acceptable chemical and physical limits for the planned duration of the proposed clinical investigation. For Phase I clinical trials, it should be confirmed that an ongoing stability program will be carried out with the relevant batch(es) and that, prior to the start of the clinical trial, at least studies under accelerated and long-term storage conditions will have been initiated (CHMP, 2006). Stability data should be generated in the proposed container/closure system outlined in 3.2.P.7 (Section 7.5.4). The actual amount of data submitted will depend upon the duration of the proposed clinical study. Therefore, if the sponsor is planning on running a very short study, then only sufficient data needed to cover the duration of the study is required.

The shelf life of the IMP should be defined based on the stability profile of the active substance and the available data on the IMP. Extrapolation may be used, provided that stability studies are conducted in parallel to the clinical studies and throughout its entire duration. This should include the proposal for shelf-life extension, defining the criteria based on which the sponsor will extend the shelf life during an ongoing study. A stability commitment should be provided (CHMP, 2006). Typically, a 12-month RED with a 30°C upper temperature limit can be applied for a "very stable formulation"[1] based on 1 month at $40 \pm 2°C/75\%$ RH $\pm 5\%$ RH open dish (CPMP, 2004). Some EU countries (i.e., United Kingdom,

---

[1] A very stable formulation is defined as a formulation that has demonstrated no change in quality of the product (i.e., no degradation is observed greater than the reporting threshold) after 1 month of open-dish storage at the accelerated conditions of 40°C/75% RH (CPMP, 2004).

Finland, Germany) and other rest of the world countries such as Malaysia and South Korea have issued guidance for RED that is more restrictive than is typically accepted by other regulatory agencies. For these countries initially, only a 6-month initial RED period can be assigned based on 1-month data.

Ideally, the filed stability data should be generated on the actual DP supplies going into the clinic. However, timeline considerations may lead to the use of a representative and closely related batch (i.e., probe batch) rather than the final clinical batch in the filing. This approach is widely accepted by the regulators, but there needs to be a commitment to also pursue stability studies on DP supplies going into the clinic. This data can then be added to the filing during routine CMC updates, for example, when updating a RED.

For supply chain efficiencies, the clinical supply and operations team may want to provide a bulk supply of DP to the clinic, which can then be repackaged into individual patient packs or dispensed on the day of dosing. If this is the case, then it's important to understand exactly how the DP will be handled at the clinic to ensure appropriate stability data is included in the filing. For example, if all of the stability data has been generated on DP contained in induction-sealed high-density polyethylene (HDPE) bottles, then repacking of these supplies into non-induction-sealed bottles will not be supported. An additional stability study should be run at the projected storage conditions on DP in non-induction-sealed packages and included in the filing. This type of data is sometimes known as "in-use" stability data.

The term "in-use" stability also applies to OSF DP, which has been manufactured at the clinic. Stability data should be generated for the entire "in-use" period of the OSF. For example, if the manufacturing process consists of the dilution of DS in a dosing vehicle contained within a glass bottle and subsequent drawing into a syringe for dosing, then stability data should be generated for the entire time the formulation is kept in the bottle and syringe prior to dosing. If, for example, the OSF is only stable for 4 h in a syringe due to sedimentation, then it must be dosed no more than 4 h after manufacture.

When generating stability data on DP formulations for a planned clinical study, it is worth considering whether there is a requirement to generate data for all of the formulations, since in some cases it may be appropriate to apply a bracketing design (ICH, 2002b). The bracketing approach is the design of a stability schedule that only samples on the extremes of certain design factors (e.g., strength, container size, and/or fill). The main assumption in this type of design is that the stability of any intermediate levels is represented by the stability of the extremes tested. If the clinical protocol is planned for dosing of 10, 50, and 100 mg potencies of the same formulation in the study, then it is feasible not to provide stability data on the 50 mg potency in a bracketing design. This approach can save resource if applied in the appropriate manner.

When designing stability studies for more complex formulations such as liquid-filled capsules, amorphous dispersions, and formulations containing salts, the project team may consider applying more frequent physical assessment in the probe stability protocol and applying a more conservative RED than what is justified by chemical stability measurements.

7.5 Information Needed in Preparation of Regulatory Submission

When the early development team have agreed upon the clinical trial design and selected a formulation approach, the sponsors must seek and obtain regulatory approval from the country in which they want to run their clinical trials before they can start. For example, if a sponsor wants to conduct a clinical trial in the United Kingdom, they must firstly obtain the appropriate approvals from the MHRA who are the competent authority in the United Kingdom. The regulatory approval process is an essential part of drug development, especially for FIH studies where there is no previous human clinical data, to ensure that clinical studies are conducted in a manner that minimizes risk to healthy volunteers and/or patients. The sponsor must provide data to the competent authority that justifies the type of clinical trial design chosen and DP formulations selected for the study. The data provided for early clinical studies of a new drug will focus on its safety with limited information provided about the formulation and its characterization (FDA, 1995).

7.5.1 Investigational New Drug and Clinical Trial Applications

Regulatory approval is requested through the IND format in the United States or CTA outside the United States. The IND or CTA is a request to the competent regulatory authority to allow administration of an investigational drug to humans. When the application is approved, it becomes the authorization to run the clinical trial. In the EU written approval is required from the competent authority in order to commence the clinical trial. Approval timelines vary from country to country but are typically in the range of 2–8 weeks for FIH studies. Conversely, written approval is not required in the United States, and the sponsor may commence the clinical trial 30 days after the FDA received the IND application. The clinical trial cannot be started if the FDA notify the sponsor that the IND has been placed on clinical hold.

The sponsor also needs to apply for, and receive, ethics approval before the trial can be commenced. Ethics approval can be obtained at the same time or after the CTA submission and is provided by an IRB in the United States or an IEC in the EU. It is important to note that the clinical trial site cannot screen for volunteers and/or patients or receive any formulations related to the specific study until the required approvals have been received from the regulatory authority.

The CTA must contain information on three broad areas, namely, animal pharmacology and toxicology studies conducted during the preclinical stage of development, DS and DP manufacturing (CMC), and clinical trial protocols and investigator. The preclinical animal pharmacology and toxicology data enables the regulator to assess whether a DP formulation containing a new API is reasonably safe for initial testing in humans. Ensuring the safety and rights of human subjects is the primary objective when regulators review a Phase I IND or CTA. The CMC section contains information on the composition, manufacturer, stability, and controls used for manufacturing the DS and the DP. It is used to assess whether the company can adequately produce and supply consistent batches of the drug. The clinical trial protocol is assessed to determine whether the initial phase trials will expose subjects to unnecessary risks. This section also contains information on the qualifications of clinical investigators—professionals (generally physicians) who oversee the administration of the experimental compound—to assess whether they are qualified to fulfill their clinical trial duties.

Tables 7.11 and 7.12 list the types of documents and content that must be included in a CTA or IND to conduct an FIH Phase I clinical trial in the United Kingdom and United States, respectively (CFR, 2016; FDA, 1995; MHRA, 2014a). In addition to the three main sections described previously, it can be seen that administrative information relating to the clinical trial application must also be included. The aim of the rest of this section is to outline general requirements for the CMC section of an Investigational Medicinal Product Dossier (IMPD) (within CTA submission) or IND submission. Other components such as the Investigator's Brochure and clinical trial protocol are not discussed.

7.5.2 Investigational Medicinal Product Dossier (IMPD) and Chemistry, Manufacturing, and Controls (CMC)

Pharmaceutical information relating to the DS and DP should be submitted to assure their proper identification, quality, purity, stability, and strength. This information is included in the IMPD or CMC section of the regulatory filing. Many guidance documents exist that readers can refer to gain an understanding of the contents and format of the IMPD/CMC section of a regulatory filing (CFR, 2016; European Commission, 2006; FDA, 1995; ICH, 2002a; MHRA, 2014a). However, some of these documents can be difficult to interpret, especially with respect to the basic contents required for a successful FIH CMC submission. Thus, the aim of this section is to provide the reader with an overview of the general CMC content and format requirements for a regulatory submission for an FIH clinical study. It is not, however, intended to provide a comprehensive description of all the requirements, since there are many country-to-country and product-to-product variations, which are beyond the scope of this chapter.

Typically, more CMC information is provided to the regulatory authorities as drug development progresses, and modifications are made to the method of preparation of the new DS and DP. Therefore, it is generally the case that when

Table 7.11 Example documentation to be submitted to the MHRA for CTA authorization for an IMPD.

Document	Description
Covering letter	Application covering letter describing the contents of the CTA. When applicable, the subject line should state that the submission is for a Phase I trial and is eligible for a shortened assessment time
EudraCT form (European Commission, 2010b)	EudraCT is a European database of clinical trial information, established by the European Medicinal Agency (EMA) for the primary purpose of providing an overview of all clinical trials conducted in the European Community. The EudraCT form is required for submission in the majority of European countries as part of a valid request for a clinical trial authorization
Clinical trial application form	Application form describing key details of the study
Clinical trial protocol	Protocol describing trial design, objectives, background and rationale, methods, procedures, statistical analysis plan, IMP details (Inc. labelling, packaging, and storage), and administrative details
Investigator's Brochure (IB)	The IB is a compilation of the clinical and preclinical data on the investigational product(s) that are relevant to the study of the product(s) in human subjects
Investigational Medicinal Product Dossier (IMPD)	A product-specific dossier that includes summaries of the pharmaceutical, nonclinical, and clinical data supporting investigational use of any IMP in humans, including reference products and placebos
Manufacturer's authorization	License from the competent authority authorizing investigational medicinal product manufacture
Qualified person declaration (required if DP or DP intermediate is manufactured outside the EU)	A document that certifies the non-EU site involved in DP manufacturing, packaging, labelling, and testing (release and stability) of the investigational medicinal products (that will be used in support of clinical trials in the EU) operates according to standards that are at least equivalent to those of the EU
Example of IMP label (or justification for its absence)	Label that is to be applied to each package containing the IMP. Labels should be in accordance with Annex 13 (European Commission, 2010a, b)

Note: A proof of payment and any previous scientific advice from the competent authority must also be included with the application.

using FFP formulations in Phase I clinical studies, the emphasis in the CMC submission is on providing information that will allow evaluation of the subject safety in the proposed study. The identification of a safety concern or insufficient data to make an evaluation of safety is the only basis for a clinical hold for an IND submission based on the CMC section. The FDA have cited some reasons

Table 7.12 Summary of IND content for submission to FDA (FDA, 1995).

Document	Description
Covering letter	Application covering letter describing (i) the details of the sponsor, (ii) date of application, (iii) name of Investigational New Drug, (iv) phase of clinical study, (v) a commitment not to begin clinical investigations until an IND covering the investigations is in effect, (vi) a commitment of compliance with an IRB, (vii) details of clinical monitor, (viii) details of person(s) responsible for safety monitoring, and (ix) signature of the sponsor
Introductory statement and general investigational plan	A brief introduction giving the name of the drug and all active ingredients, the drug's pharmacological class, the structural formula of the drug, the formulation of the dosage form(s) to be used, the route of administration, and the broad objectives and planned duration of the proposed clinical investigation(s). Also, a brief description of the overall plan for investigating the DP for the following year should be included (this is also known as a general investigational plan)
Investigator's Brochure (IB)	The IB is a compilation of the clinical and nonclinical data on the investigational product(s) that are relevant to their study in human subjects. It should generally include a brief description of the DS and the formulation, including the structural formula, summary of the pharmacological and toxicological effects of the drug in animals, summary of the pharmacokinetics and biological disposition of the drug in animals, and a summary of possible risks and side effects to be anticipated
Protocols	Phase I protocols should be directed primarily at providing an outline of the investigation—an estimate of the number of patients to be involved, a description of safety exclusions, and a description of the dosing plan including duration, dose, or method to be used in determining dose—and should specify in detail only those elements of the study that are critical to safety, such as necessary monitoring of vital signs and blood chemistries
Chemistry, manufacturing, and control information	Section describing the composition, manufacture, and control of the DS, DP, and placebo (if used in a controlled clinical trial). Environmental analysis requirements and a copy of all labels to be provided to each investigator should also be included. The emphasis in an initial Phase I submission should generally be placed on the identification and control of the raw materials and the new DS. Final specifications for the DS and DP are not expected until the end of the investigational process
Pharmacology and toxicology information	Include adequate information about pharmacological and toxicological studies of the drug involving laboratory animals or *in vitro*, on the basis of which the sponsor has concluded that it is reasonably safe to conduct the proposed clinical investigations
Previous human experience with the investigational drug	If the investigational drug has been previously dosed in humans, a summary report of the findings should be provided. The application should explicitly state if there has been no previous human experience with the drug
Additional information	This includes any other relevant information such as drug dependence and abuse potential, radioactive drugs, and pediatric studies

Note: A table of contents should also be included after the covering letter.

when they may have concern on the CMC sections including (i) a formulation that cannot remain chemically stable throughout the testing program proposed, (ii) a formulation made with unknown or impure components, (iii) a product possessing chemical structures of known or highly likely toxicity, or (iv) a product with an impurity profile indicative of a potential health hazard or an impurity profile insufficiently defined to assess a potential health hazard (FDA, 1995).

7.5.3 The Common Technical Document

The IMPD is typically organized using the Common Technical Document (CTD) format (European Commission, 2006; ICH, 2002a). The CTD (Figure 7.8) is an internationally agreed format for the preparation of applications to be submitted to regulatory authorities in the three ICH regions of Europe, United States, and Japan. It is intended to save time and resources and to facilitate regulatory review and communication. The CTD gives no information about the content of a dossier and does not indicate which studies and data are required for a successful approval. As discussed previously, regional requirements may affect the content of the dossier submitted in each region; therefore, the dossier will not necessarily be identical for all regions.

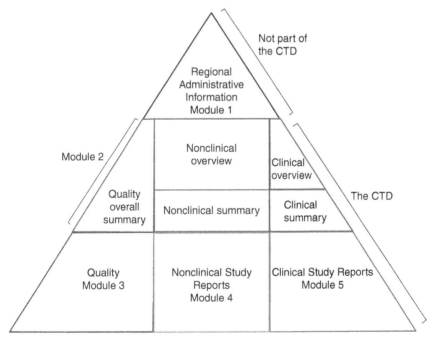

Figure 7.8 The Common Technical Document (CTD) triangle. The CTD is organized into five modules. Module 1 is region specific and modules 2–5 are intended to be common for all regions. A summary of the CMC information is contained within quality of overall summary Module 2, and more detailed information can be found within quality Module 3.

The quality section of the CTD provides a harmonized structure and format for presenting CMC information in a registration dossier. A summary of the quality information is contained within Module 2, whereas the detailed information is contained within Module 3. Module 3 mainly consists of the DS (Section 3.2.S) and DP (Section 3.2.P) sections. An additional appendices (Section 3.2.A) and regional information section (Section 3.2.R) may be included, if required. The DS and DP information should follow the format laid out in Tables 7.13 and 7.14, respectively. Each unique formulation to be investigated in the clinical study will have a separate DP and DS section (if a different DS is being utilized). It is feasible to include different potencies of the same formulation in a single filing. For example, if the study protocol dictates that both a suspension and tablet formulation will be dosed, then these

Table 7.13 Format and content of the Common Technical Document Module 3: Drug Substance.

3.2.S Drug Substance (Name and Manufacturer)
3.2.S.1 General Information
3.2.S.1.1 Nomenclature
3.2.S.1.2 Structure
3.2.S.1.3 General Properties
3.2.S.2 Manufacture
3.2.S.2.1 Manufacturer(s)
3.2.S.2.2 Description of Manufacturing Process and Process Controls
3.2.S.2.3 Control of Materials
3.2.S.2.4 Controls of Critical Steps and Intermediates
3.2.S.2.5 Process Validation and/or Evaluation
3.2.S.2.6 Manufacturing Process Development
3.2.S.3 Characterization
3.2.S.3.1 Elucidation of Structure and Other Characteristics
3.2.S.3.2 Impurities
3.2.S.4 Control of Drug Substance
3.2.S.4.1 Specification
3.2.S.4.2 Analytical Procedures
3.2.S.4.3 Validation of Analytical Procedures
3.2.S.4.4 Batch Analyses
3.2.S.4.5 Justification of Specification
3.2.S.5 Reference Standards or Materials
3.2.S.6 Container Closure Systems
3.2.S.7 Stability
3.2.S.7.1 Stability Summary and Conclusions
3.2.S.7.2 Post-Approval Stability Protocol and Stability Commitment
3.2.S.7.3 Stability Data

Table 7.14 Format and content of the Common Technical Document Module 3: Drug Product.

3.2.P	Drug Product (Name, Dosage Form, and Manufacturer)
	3.2.P.1 Description and Composition of the Drug Product
	3.2.P.2 Pharmaceutical Development
	3.2.P.3 Manufacture
	3.2.P.3.1 Manufacturer(s)
	3.2.P.3.2 Batch Formula
	3.2.P.3.3 Description of Manufacturing Process and Process Controls
	3.2.P.3.4 Controls of Critical Steps and Intermediates
	3.2.P.3.5 Process Validation and/or Evaluation
	3.2.P.4 Control of Excipients (Name)
	3.2.P.4.1 Specification(s)
	3.2.P.4.2 Analytical Procedures
	3.2.P.4.3 Validation of Analytical Procedures
	3.2.P.4.4 Justification of Specifications
	3.2.P.4.5 Excipients of Human or Animal Origin
	3.2.P.4.6 Novel Excipients
	3.2.P.5 Control of Drug Product
	3.2.P.5.1 Specification(s)
	3.2.P.5.2 Analytical Procedures
	3.2.P.5.3 Validation of Analytical Procedures
	3.2.P.5.4 Batch Analyses
	3.2.P.5.5 Characterization of Impurities
	3.2.P.5.6 Justification of Specification(s)
	3.2.P.6 Reference Standards or Materials
	3.2.P.7 Container Closure System
	3.2.P.8 Stability
	3.2.P.8.1 Stability Summary and Conclusion
	3.2.P.8.2 Post-Approval Stability Protocol and Stability Commitment
	3.2.P.8.3 Stability Data

will require separate DP sections in the filing. However, 50 and 100 mg tablets using the same DS can be included within the same filing. If the study also includes a placebo arm, then another separate DP filing must be submitted as part of the CTA or IND.

Most companies will typically file the CMC sections for the proposed clinical trial when they have gathered all the relevant data. In certain instances, where all the data may not be available, for example, clinical stability data, then relevant (probe) batch analysis data will be sufficient as long as the CMC sections are updated when the clinical data becomes available.

7.5.4 Notes on the Content of the CMC Section

3.2.S *Drug Substance Information*

    3.2.S.1 *General Information*

        3.2.S.1.1 *Nomenclature*

        The nomenclature of the DS should be provided. This is typically the chemical name and the company/laboratory code.

        3.2.S.1.2 *Structure*

        The structural formula, molecular formula, and relative molecular mass should be described. Also, reference to the relative and absolute stereochemistry should be made.

        3.2.S.1.3 *General Properties*

        A list of the API physiochemical and other relevant properties should be provided. This can take the form of a table and will normally include the description, hygroscopicity, optical rotation, solubility (organic and aqueous), pH, dissociation constant, thermal behavior, and crystallinity and polymorphism of the API (ICH, 1999).

    3.2.S.2 *Manufacture*

        3.2.S.2.1 *Manufacturers*

        Each manufacturer, including contractors and testing labs, and each production site involved in the manufacture of the DS should be listed.

        3.2.S.2.2 *Description of Manufacturing Process and Process Controls*

            3.2.S.2.2.1 *Process Flow Diagram*

            A flow diagram of the manufacturing process for the API should be provided. This should include the intermediates, solvents, catalysts, and critical reagents used. Some countries, for example, Germany, require all the reagents to be listed and not just those deemed critical. Chemical structures of starting materials, intermediates, reagents, and DS reflecting stereochemistry should be provided. Typically, only the GMP steps will be included in this section.

            3.2.S.2.2.2 *Description of Process: General Description*

            A brief narrative summary of the GMP manufacturing process from DS starting materials(s) to the final isolated DS should be included. The summary should include intermediates, solvents, catalysts, and critical reagents. The materials described in this process description must match those depicted in the process flow diagram. If the material has been milled, then a statement to this effect should also be added.

3.2.S.2.3 *Control of Materials*

3.2.S.2.3.1 *Starting Materials*

In this section, appropriate controls for the starting materials and intermediates should be described. The starting materials and intermediates should be listed and identify where each material is used in the process.

3.2.S.2.3.2 *Critical Reagents and Solvents*

Typically, a list of the full chemical names of the solvents, catalysts, and critical reagents used in the reactions for the GMP steps will be provided in a table format. All reagents and solvents listed should appear in both the process flow diagram and the process description.

3.2.S.2.4 *Controls of Critical Steps and Intermediates*

Generally, this information is not required for FIH studies.

3.2.S.2.5 *Process Validation and/or Evaluation*

Generally, this information is not required for FIH studies.

3.2.S.2.6 *Manufacturing Process Development*

A description of the current manufacturing process should be outlined in this section. Note: A reference can be made to Section 3.2.S.2.2. A brief summary comparing the DS used in the preclinical toxicology studies with that used in the CTA/IND filing should also be provided. If there are any differences, these should be described, for example, if a new salt form was manufactured for the GMP delivery.

3.2.S.3 *Characterization*

3.2.S.3.1 *Elucidation of Structure and Other Characteristics*

The DS structure based on, for example, synthetic route and spectral analyses should be confirmed and provided. Information such as the potential for isomerism, the identification of stereochemistry, or the potential for forming polymorphs should also be included.

Spectroscopic methods such as ultraviolet/visible spectrophotometry, infrared transmittance, proton magnetic resonance, carbon-13 magnetic resonance, and mass spectrophotometry can be used to elucidate the structure of the DS.

3.2.S.3.2 *Impurities*

Details of process solvents and inorganic and mutagenic impurities should be provided.

3.2.S.4 *Control of Drug Substance*

3.2.S.4.1 *Specification*

A specification for the DS should be provided. This should include the analytical test, acceptance criteria, and method used. Using assay as a test example, the acceptance criteria could be 97.0–103%, and high-performance liquid chromatography (HPLC) could be the method used (ICH, 1999). A note should also be made regarding the impurity profile of the DS with reference to ICH Q3A (ICH, 2006a).

3.2.S.4.2 *Analytical Procedures*

A brief description of the non-compendial analytical methods developed for the DS should be provided. For example, this could include an assay by titration, assay and related compounds (HPLC), and residual solvents (gas chromatography) (ICH, 2005).

3.2.S.4.3 *Validation of Analytical Procedures*

Generally, this information is not required for FIH studies. However, this is dependent on local regulatory authority guidance and should be confirmed via appropriate local guidances.

3.2.S.4.4 *Batch Analysis*

Results for batches used in the current clinical trial, in the preclinical animal toxicology studies, and, where applicable, for all batches used in previous clinical trials should be supplied.

3.2.S.4.5 *Justification of Specification*

A justification for the DS specification should be provided.

3.2.S.5 *Reference Standard or Materials*

Normally, a primary reference standard will not be established for FIH studies. A well-characterized lot of DS can be used as a working standard (ICH, 1999).

3.2.S.6 *Container Closure System*

A brief description of the container closure system should be provided including the identity of the materials of construction of each packaging component.

Table 7.15 Example of a drug substance stability protocol.

Condition	Time (months)						
	0	1	3	6	12	24	36
25°C/60% RH	X	—	—	—	X	X	X
40°C/75% RH		X	X	X	—	—	—

3.2.S.7 *Stability*

   3.2.S.7.1 *Stability Summary and Conclusions*

The types of studies conducted, protocols used, and the results of the studies should be summarized. The summary should include results, for example, from forced degradation studies and stress conditions, as well as conclusions with respect to storage conditions and retest date or shelf life, as appropriate (ICH, 1996b, 2003).

   3.2.S.7.2 *Stability Protocol*

This section should describe the post-approval stability protocol and stability commitment. It's common to see a stability protocol utilizing accelerated conditions, for example, 40°C/75% RH for the early time points (e.g., up to 6 months) and then switch to real-time stability testing, for example, 25°C/65% RH for the longer time points (6 months–3 years). Table 7.15 shows an example DS stability protocol.

   3.2.S.7.3 *Stability Data*

A tabulated summary of the API stability data should be provided. (Note: Graphical and narrative summaries are also acceptable.) For early phase filings this may include only 1 month accelerated stability data. Reference ICH Guidelines (ICH, 1996b, 2003, 2005).

3.2.P *Drug Product Information*

   3.2.P.1 *Description/Composition of the Drug Product*

In this section, compositional details of each strength of the specific formulation to be studied in the clinical trial should be listed. These details are commonly listed in a table format within this section of the CTD (Table 7.16). For each active ingredient and excipient, the following details should be listed:

- Details of the quality standard in which the material is tested/ released against, for example, a compendial monograph.

Table 7.16 Drug product composition.

Components	Quality standard	Function	Unit strength (mg/capsule)	
			0.5 mg	75 mg
API	In-house	Active	0.500	75.00
Microcrystalline cellulose	USP	Diluent	28.70	63.30
Lactose monohydrate	USP	Diluent	114.8	253.2
Croscarmellose sodium	USP	Disintegrant	4.500	13.50
Magnesium stearate	USP	Lubricant	1.500	4.500
Capsule fill weight			150.0	450.0
Size #3, white opaque hard gelatin capsule	—	—	One capsule	—
Size #00, white opaque hard gelatin capsule	—	—	—	One capsule

- Function—role material play in the formulation, for example, magnesium stearate is a lubricant; microcrystalline cellulose and lactose monohydrate are diluents.
- Unit formula for each strength of formulation to be tested, including total fill weight.

3.2.P.2 *Pharmaceutical Development*

In this section, a brief description of the pharmaceutical development activities leading up to the selection of the clinical formulation outlined in 3.2.P.1 should be given. If this is the first CTA filing, a simple statement referring to the composition in 3.2.P.1 and the formula in 3.2.P.3 should suffice as a description. If the API has already been investigated as a different formulation in another CTA or IND, for example, current application is for a tablet formulation, whereas a solution OSF was previously been filed, then a more detailed description is required. In this case, a brief description of the solution OSF composition (qualitative and not quantitative) and process should be given, followed by reference to the new tablet composition and process. A rationale for the formulation switch should also be provided.

When the FIH approach utilizes an OSF manufactured at the clinic, for example, DS suspended in a dosing vehicle, then dose accuracy studies should be performed and documented in this section. The aim of these studies is to demonstrate that the preparation and dispensing process will achieve the targeted dose, that is, within the range of 90.0–110.0% of the required clinical dose.

3.2.P.3 *Manufacture*

3.2.P.3.1 *Manufacturer(s)*

This section should contain a list of the manufacturers involved in the product supply chain. In this case, manufacturers are responsible for manufacturing, packaging, and testing the finished product. The list should include the name, address, and responsibility of each manufacturer (including contractors).

3.2.P.3.2 *Batch Formula*

A batch formula should be provided, which includes a list of all the components used in the manufacturing process and their amounts (on a per batch basis). If multiple potencies are being used in the study, these can all be listed in the same table.

3.2.P.3.3 *Description of Manufacturing Process and Process Controls*

A flow diagram of the successive steps, indicating the components used for each step including any relevant in-process controls, should be provided. In addition, a brief narrative description of the manufacturing process should be included. Nonstandard manufacturing processes or new technologies and new packaging processes should be described in more detail. An example flow diagram for a simple direct encapsulated dry-filled blend formulation (i.e., DFC) is shown in Figure 7.9.

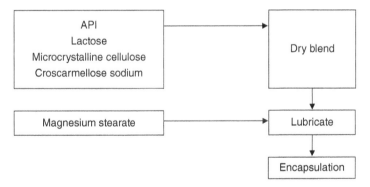

Figure 7.9 Flow diagram describing the manufacturing process for the drug product shown in Table 7.15.

3.2.P.3.4 *Controls of Critical Steps and Intermediates*
This section is generally not required for FIH/early phase studies, unless a nonstandard manufacturing process(es) or sterile products are being used for the clinical study.

3.2.P.3.5 *Process Validation and/or Evaluation*
This information is typically not required for FIH studies.

3.2.P.4 *Control of Excipients*
This section should detail the specifications that the excipients used in the clinical formulation are tested. If all of the excipients are tested to a compendial specification, for example, the USP, then a simple statement stating this fact should be sufficient.

In cases where a premixed blend of compendial excipients is also used in the formulation, for example, a prefabricated dry mix such as Opadry® for film coating, then details of each of the excipients in the mix should be described and a representative CoA included in the document.

The inclusion of novel excipients in the formulation calls for more information to be provided so that the reviewer can assess implications of dosing in patients. In some cases, a toxicology package similar to that provided for an API may need to be included. Alternatively, for a US regulatory filing, reference to a Drug Master File (DMF) can be provided, if one is available. A letter of authorization to reference the DMF will also be required.

Details of the analytical procedures used to test the excipients against the specifications should be described. If all the excipients are tested using compendial methods, a simple statement to this effect will be sufficient, for example, "excipients are tested as per the compendia or accepted on supplier's CoA." If the excipient(s) are tested using non-compendial methods, these should be outlined in detail.

Information on the validation of analytical procedures should also be provided. In the case of compendial methods, these will be considered to be sufficiently validated. A note should also be made relating to the justification of the specification.

Any excipients included in the formulation that are manufactured from an animal or human origin should also be listed in this section. Lactose and gelatin are common pharmaceutical excipients that are derived from animal origins. The sponsor should clearly demonstrate conformance with relevant guidelines, the current European Committee for Proprietary Medicinal Product

(CPMP)–Committee for Veterinary Medicinal Product (CVMP) Note for Guidance on Minimizing the Risk of Transmitting Animal Spongiform Encephalopathy Agents via Human and Veterinary Medicinal Products (EMEA/410/01).

3.2.P.5 *Control of Drug Product*

3.2.P.5.1 *Specification(s)*

In this section, a table listing the key DP specifications should be inserted. The table should include the name of the test, the procedure/method, and the acceptance criteria. Reference ICH Guidelines (ICH, 1999, 2006b).

3.2.P.5.2 *Analytical Procedures*

This section should provide a brief description of the non-compendial analytical methods developed for the DP. Reference ICH Guideline (ICH, 2005).

3.2.P.5.2.1 *Identification*

This method is typically similar to the method used for assay and degradation analysis (Section 3.2.P.5.2.2). Where the same method is used, a reference to Section 3.2.P.5.2.2 should be sufficient. If a different method is used, then a tabulated summary of the analytical method conditions should be added. For HPLC methods, the table should include the column name, column temperature, flow rate, injection volume, run time, and mobile phase(s) and whether the method is isocratic or gradient.

3.2.P.5.2.2 *Assay and Degradation Products (HPLC)*

A tabulated summary of the HPLC conditions should be listed including the column name, column temperature, flow rate, injection volume, run time, and mobile phase(s) and whether the method is isocratic or gradient.

3.2.P.5.2.3 *Uniformity of Dosage Units*

A tabulated summary of the analytical method conditions should be listed. For HPLC this would include the column name, column temperature, flow rate, injection volume, run time, and mobile phase(s) and whether an isocratic or gradient method was used.

3.2.P.5.3 *Validation of Analytical Procedures*
Validation information and data on the analytical procedures used for the testing of the DP should be provided. This should include information on the specificity, linearity and range, precision, accuracy, limit of quantification, and solution stability of the methods. Reference ICH Guideline (ICH, 2005).

3.2.P.5.4 *Batch Analysis*
A tabulated description (manufacturing site, batch size, batch/lot number, date of manufacture, DS batch used) and data should be provided for each formulation included in the regulatory application. Data should be provided for each of the tests mentioned in 3.2.P.5.1. In some cases, it may be beneficial to file representative batch (i.e., probe) analysis data. For example, when development programs are being executed under aggressive timelines, the clinical batch has not been manufactured prior to the CTA or IND being submitted to the regulatory authority. A representative (i.e., probe) batch is typically manufactured at a smaller scale than the clinical batch; however, the process and formulation should be representative of the batch, which will be used in the clinical study. Reference ICH Guidelines (ICH, 1999, 2005, 2006b).

3.2.P.5.5 *Characterization of Impurities*
Information on the characterization of impurities should be provided. The analytical procedures for evaluating degradation products in the DP should be described in Section 3.2.P.5.2. The data on the degradation products that are found in the DP batches will be presented in Section 3.2.P.5.4. The impurities and degradation products observed in the batches should either be qualified by a preclinical animal toxicology study or be below the qualification thresholds outlined in ICH Q3B Guideline on Impurities in New Drug Products (ICH, 2006b).

3.2.P.5.6 *Justification of Specification*
Justification for the proposed DP specifications should be provided, for example, if the specification has been set against ICH guidances and compendial requirements (ICH, 1999, 2006b).

3.2.P.6 *Reference Standard or Materials*

A reference standard, or reference material, is a substance pre-pared for use as the standard in an assay, identification, or purity test. For FIH regulatory applications, a primary reference standard will generally not be established. When this is the case, a state-ment should be added to this section stating that a previously released batch of DS is used as the reference standard. Reference ICH Guidelines (ICH, 1999).

3.2.P.7 *Container Closure System*

The packaging used for the clinical supplies should be described in this section. For example, if the DP will be supplied to the clinic in induction-sealed HDPE bottles with child-resistant caps, all this information including size of bottles and caps should be stated. If the packaged product requires desiccation due to moisture-related stability issues, then information on the num-ber and type of desiccants per pack should be documented.

3.2.P.8 *Stability*

3.2.P.8.1 *Stability Summary and Conclusion*

A brief summary of the studies undertaken (conditions, batches, analytical procedures) and a brief discussion of the results and conclusions of the stability studies and analysis of data should be included. Conclusions with respect to storage conditions and shelf life and, if applica-ble, in-use storage conditions and shelf life should be given. Reference ICH Guidelines (ICH, 1996b, 1999, 2003).

3.2.P.8.2 *Stability Protocol and Stability Commitments*

A description of the post-approval stability protocol and stability commitment should be provided. This can take the form of a table outlining the proposed storage conditions and time points for analysis (ICH, 2003).

3.2.P.8.3 *Stability Data*

Results of the stability studies should be presented in an appropriate format (e.g., tabular, graphical, and nar-rative). Information on the analytical procedures used to generate the data and validation of these procedures should be included. For FIH studies, it's common to file 1-month accelerated probe data to shorten the time to clinical study start. In this case, the filing would need to be updated with clinical stability data through-out the duration of the study and until the longest pro-jected time point. Reference ICH Guidelines (ICH, 1996b, 2003, 2005).

7.5.5 Flexible Formulation Design Space and Clinical Protocols

In recent years there has been a trend toward introducing flexibility within the CMC section of the regulatory filing. Rather than describing each specific formulation to be dosed in the clinical study in 3.2.P.1, it is instead feasible to detail a formulation design space. Thinking of the formulation and/or process in terms of a design space avoids the constraint of predetermination and premanufacture of fixed quantitative DP compositions. Instead, a bracketing strategy can be adopted to describe quantitative, continuous ranges in excipient levels within which an optimal DP formulation composition is anticipated to lie (McDermott and Scholes, 2015). In this case, representative batch analysis and stability data at the extremes of the design space should be included in the filing to provide assurance of the DP quality of any discrete formulation composition within the defined range.

The formulation design space submission can also be coupled with fast GMP manufacture, QC testing, QP release, clinical dosing, and bioanalysis to enable "real-time" decision making in the clinic. The beauty of this approach is the flexibility to adjust the formulation composition within the clinical study based on emerging human safety, pharmacokinetic, and/or pharmacodynamic data. Studies can be run with one or more formulation/process variables creating a multidimensional design space.

The formulation design space concept can be thought of in terms of pharmaceutical development within a QbD framework (ICH, 2009c). To these authors' knowledge the formulation design space concept has only achieved regulatory acceptance from the UK MHRA. However, it seems feasible that this concept could be applied to other regions.

The formulation design space approach has mostly been employed to optimize the DP composition of an NCE either before committing to pivotal Phase II/III studies or as part of a product value enhancement (PVE) strategy to switch delivery routes or improve the dosage regimen to aid patient compliance/medicines adherence. Some examples include the establishment of a level A *in vitro–in vivo* correlation (IVIVC) for a modified release dosage form (Kane *et al.*, 2015) and the optimization of an enabled formulation (McDermott *et al.*, 2014). In the enabled formulation development study, the authors compared formulations of IDX-719 containing two different polymers and for each formulation a two-dimensional design space with surfactant and acidic modifier levels. IDX-719 is a BCS Class IV molecule where no predictive *in vitro* or preclinical animal model data was available to support clinical formulation development and selection. Conducting a RapidFACT program enabled an acceleration of program timelines through the rapid flexible comparison of the different prototypes in the clinic.

The formulation design space CMC filing has also been utilized within adaptive Phase I clinical studies (described in Section 7.3.2), which can enable rapid assessment of multiple formulations, dosing regimens, and (informal) food effect within a single protocol (Connor *et al.*, 2013). Efficient, just-in-time GMP manufacturing of DP also enables rapid study start by minimizing DP stability timeframe requirements (Hunt *et al.*, 2015; Millington *et al.*, 2015).

7.6 Conclusions

Effectively planning of the activities to support the FIH study is critical to ensure the safety and well-being of study volunteers, as well as maximize efficiency and scientific validity of the study while reducing resource investment in the clinical trial process.

The initial section of the chapter has discussed the most important factors to consider when planning, designing, and conducting an FIH clinical trial, ranging from subject and trial site selection, study design and data analysis methods. As discussed, the choice of trial design (e.g., SAD, MAD, or oncology-specific designs) heavily depends on the outcomes of the study, the type of compound (e.g., long half-life, stability requirements, site capabilities for reconstitution), funding, and other capabilities (e.g., data analysis). As well as the traditional designs, this section has also covered the newer "multipart flexible" trial designs, which enable the use of "real-time" data to guide dosing and formulation decisions up front. Overall, effectively planning and designing an FIH clinical trial will enable the production of high quality data to achieve trial outcomes while maximizing efficiency. Most importantly, this will also help to ensure the safety of study volunteers and facilitate the progression of the compound into later phase trials.

Subsequently, the chapter has addressed the development and manufacturing strategy for the formulation supplied to the FIH study. The concepts of FFP and OSF as strategies to reduce the time to enter the clinical trial phase have been introduced, as well as the role of solubility and trial design in driving the development effort. Various formulation approaches to supply the clinical study have been presented including ready-to-use solutions or suspensions, API in bottle, API in capsule, and conventional capsules and tablets. Selecting the right formulation approach can help to reduce the time to dosing and enables the quicker production of clinical data. The formulation used should be reviewed on a case-by-case basis, taking into account the drug properties, cost of manufacture, and the overall study aims.

An overview of the analytical activities that take place during FFP formulation development was also presented. The analytical section focused on product specification setting, analytical method development, analytical method validation, batch analysis, and stability testing requirements for early phase clinical studies. It also discussed different strategies for generating data for the regulatory filing, for example, using probe batch analysis and stability data to enable an earlier regulatory submission date.

Finally, a discussion on the information required for the preparation of a regulatory submission for an FIH trial was presented. This section initially describes both CTA and IND submissions and summarizes their contents. A deeper overview of the IMPD and CMC format and content was then provided. This section is intended to give the reader an indication of the type of information that should be included within the DS and DP sections of a regulatory filing for an FIH study. The concept of a flexible formulation design space that enables real-time decision making in the clinic has also been introduced.

Abbreviations

AUC	area under the curve
ADME	absorption, distribution, metabolism and excretion
API	active pharmaceutical ingredient
ATD	accelerated titration design
BCS	biopharmaceutics classification system
BC	biocomparison studies
BE	bioequivalence studies
CHMP	Committee for Medical Product for Human Use
C_{max}	maximum plasma concentration
CMC	chemistry manufacturing and control
CPMP	Committee for Proprietary Medicinal Product
CRM	continual reassessment method
CRO	Contract Research Organization
CTA	Clinical Trial Authorisation
CTD	common technical document
CVMP	Committee for Veterinary Medicinal Product
DDI	drug–-drug interaction
DFC	dry filled capsule
DLT	dose limiting toxicity
DMF	drug master file
DP	drug product
DS	drug substance
D_o	dose number
EMA	European Medicines Agency
EWOC	escalation with overdose control
FaSSIF	Fasted State Simulated Intestinal Fluid
FDA	Food and Drug Administration
FFP	fit-for-purpose
FIH	first-in-human

GMP	good manufacturing procedures
HED	human equivalent dose
HDPE	high-density polyethylene
HME	hot melt extrusion
HPLC	high pressure liquid chromatography
ICH	International Conference on Harmonisation
IEC	Independent Research Ethics Committee
IMP	investigational medicinal product
IMPD	investigational medicinal product dossier
IND	investigational new drug
IRB	Institutional Review Board
IVIVC	*in vitro in vivo* correlation
LOQ	limit of quantitation
MABEL	minimum anticipated biological effect level
MAD	multiple ascending dose
MHRA	Medicines and Healthcare products Regulatory Agency
MIA	Manufacture Importer Authorization
MRSD	maximum recommended starting dose
MTD	maximum tolerated dose
NCE	new chemical entity
NOAEL	no observed adverse effect level
OSF	on-site formulations
PMDA	Pharmaceuticals and Medical Devices Agency
QbD	quality by design
QP	qualified person
RH	relative humidity
RSD	relative standard deviation
SAD	single ascending dose
SGF	simulated gastric fluid
SMEDDS	self-emulsifying drug delivery systems
STD	severe toxicity dose
T_{max}	time to maximum plasma concentration
USP	United States Pharmacopeia
WOCBP	Women of Childbearing Potential

Acknowledgments

The authors would like to gratefully thank Dave Storey for his invaluable inputs and helpful discussions. Cristina Dumitru and Daniel Greenwood are thanked for the literature review on FIH trial design.

References

Adkin A, Davis S, Sparrow R, Huckle P, Phillips J, Wilding I. 1995. The effects of pharmaceutical excipients on small intestinal transit. *Br J Clin Pharmacol.* 39: 381–387.

Allison M. 2012. Reinventing clinical trials. *Nat Biotechnol.* 30: 41–49.

Amidon G, Oh D, Curl R. 1993. Estimating the fraction dose absorbed from suspensions of poorly soluble compounds in humans: a mathematical model. *Pharm Res.* 10(2): 264–270.

Amidon G, Lennernäs H, Shah V, Crison J. 1995. A theoretical basis for a biopharmaceutic drug classification: the correlation of *in vitro* drug product dissolution and *in vivo* bioavailability. *Pharm Res.* 12(3): 413–420.

Association of the British Pharmaceutical Industry. 2011. First in human studies: points to consider in study placement, design and conduct. Available (Online) at: http://www.abpi.org.uk/ourwork/library/guidelines/Documents/First%20 in%20Human%20Studies.pdf (accessed November 9, 2015).

Aubry A, Sebastian D, Hobson T, Xu J Q, Rabel S, Xie M, Gray V. 2000. In-use testing of extemporaneously prepared suspensions of second generation non-nucleoside reversed transcriptase inhibitors in support of phase I clinical studies. *J Pharm Biomed Anal.* 23: 535–542.

Babb J, Rogatko A, Zacks S. 1998. Cancer phase I clinical trials: efficient dose escalation with overdose control. *Stat Med.* 17: 1103–1120.

Becquemont L, Verstuyft C, Kerb R, Brinkmann U, Lebot M, Jaillon P, Funck-Brentano C. 2001. Effect of grapefruit juice on digoxin pharmacokinetics in humans. *Clin Pharmacol Ther.* 70: 311–316.

Benet L Z. 2013. The role of BCS (biopharmaceutics classification system) and BDDCS (biopharmaceutics drug disposition classification system) in drug development. *J Pharm Sci.* 102: 34–42.

Bi M, Sun C C, Alvarez F, Alvarez-Nunez F. 2010. The manufacture of low-dose oral solid dosage form to support early clinical studies using an automated micro-filing system. *AAPS PharmSciTech.* 12: 88–95.

Breitenbach J. 2002. Melt extrusion: from process to drug delivery technology. *Eur J Pharm Biopharm.* 54(2): 107–117.

Buckley S, Frank K, Fricker G, Brandl M. 2013. Biopharmaceutical classification of poorly soluble drugs with respect to "enabling formulations". *Eur J Pharm Biopharm.* 50: 8–16.

Buoen C, Bjerrum O, Thomsen M. 2005. How first-time-in-human studies are being performed: a survey of phase I dose-escalation trials in healthy volunteers published between 1995 and 2004. *J Clin Pharmacol.* 45: 1123–1136.

Capsugel. 2008. Xcelodose®S Product Information Leaflet. Available (Online) at: http://capsugel.com/media/library/xcelodose_s_precision_powder_micro_ dosing_system.pdf (accessed January 9, 2016).

21 CFR. 2016. Code of Federal Regulations, Title 21, Chapter, Subchapter D, Part 312, Subpart B, §312.23, IND content and format. Available (Online) at: http://www.ecfr.gov/cgi-bin/text-idx?SID=16f07178d222cf97596fa0e04d61e6b9&mc=true&node=se21.5.312_123&rgn=div8 (accessed June 20, 2016).

Charman S, Charman W, Rogge M, Wilson T, Dutko F, Pouton C. 1992. Self-emulsifying drug delivery systems: formulation and biopharmaceutic evaluation of an investigational lipophilic compound. *Pharm Res.* 9(1): 87–93.

Chatterjee A, Digumarti R, Katneni K, Upreti V, Mamidi R, Mullangi R, Surath A, Srinivas M, Uppalapati S, Jiwatani S, Srinivas N. 2005. Safety, tolerability and pharmacokinetics of a capsule formulation of DRF-1042, a novel camptothecin analog, in refractory cancer patients in a bridging phase I study. *J Clin Pharmacol.* 45: 453–460.

Committee for Medicinal Products for Human Use (CHMP). 2006. Guideline on the requirements to the chemical and pharmaceutical quality documentation concerning investigational medicinal products in clinical trials. Available (Online) at: http://ec.europa.eu/health/files/eudralex/vol-10/18540104en_en.pdf (accessed June 14, 2016).

Committee for Proprietary Medicinal Products (CPMP). 2004. Guideline on stability testing: stability testing of existing active substances and related finished products. Available (Online) at: http://www.ema.europa.eu/docs/en_GB/document_library/Scientific_guideline/2009/09/WC500003466.pdf (accessed June 14, 2016).

Connor A, Scholes P, Stevens L. 2013. Flexible approaches to incorporate formulation assessments and selection into first-in-human (FIH) safety and tolerability studies. Presented at: *American Association of Pharmaceutical Sciences (AAPS) Conference, San Antonio, TX, November 10–14, 2013.*

Dobry D, Settell S, Baumann J, Ray R, Graham L, Beyerinck R. 2009. A model-based methodology for spray-drying process development. *J Pharm Innov.* 4: 133–142.

Dresser R. 2009. First-in-human trial participants: not a vulnerable population, but vulnerable nonetheless. *J Law Med Ethics.* 37(1): 38–50.

European Commission. 2006. Volume 2B—Notice to applicants—Medicinal products for human use—Presentation and format of the dossier—Common Technical Document (CTD). Available (Online) at: http://ec.europa.eu/health/files/eudralex/vol-2/b/update_200805/ctd_05-2008_en.pdf (accessed June 13, 2016).

European Commission. 2010a. The rules governing medicinal products in the European Union. Volume 4—EU guidelines to Good Manufacturing Practice. Medicinal products for human and veterinary use. Annex 13. Investigational medicinal products. Available (Online) at: http://ec.europa.eu/health/files/eudralex/vol-4/2009_06_annex13.pdf (accessed June 13, 2016).

European Commission. 2010b. Communication from the Commission—detailed guidance on the request to the competent authorities for authorization of a clinical trial on a medicinal product for human use, the notification of substantial amendments and the declaration of the end of the trial (CT-1) (2010/C 82/01). Available (Online) at: http://ec.europa.eu/health/files/eudralex/vol-10/2010_c82_01/2010_c82_01_en.pdf (accessed June 15, 2016).

European Medicines Agency (EMA); Committee for Medicinal Products for Human Use (CHMP). 2007. Guideline on strategies to identify and mitigate risks for first-in-human clinical trials with investigational medicinal products. Available (Online) at: http://www.ema.europa.eu (accessed October 27, 2015).

European Medicines Agency (EMA); Committee for Medicinal Products for Human Use (CHMP). 2010. Guideline on the investigation of bioequivalence. Available (Online) at: http://www.ema.europa.eu/docs/en_GB/document_library/Scientific_guideline/2010/01/WC500070039.pdf (accessed October 21, 2015).

Fish D N, Vidaurri V A, Deeter R G. 1999. Stability of valacyclovir hydrochloride in extemporaneously prepared oral liquids. *Am J Health Syst Pharm.* 56: 1957–1960.

Food and Drug Administration (FDA). 1993. Guidance for Industry: guideline for the study and evaluation of gender differences in the clinical evaluation of drugs. Available (Online) at: http://www.fda.gov/downloads/ScienceResearch/SpecialTopics/WomensHealthResearch/UCM131204.pdf (accessed January 14, 2016).

Food and Drug Administration (FDA). 2002. Guidance for Industry: food-effect bioavailability and fed bioequivalence studies. Available (Online) at: http://www.fda.gov/downloads/RegulatoryInformation/Guidances/UCM126833.pdf (accessed December 8, 2015).

Food and Drug Administration (FDA). 2005. Guidance for Industry—estimating the maximum safe starting dose in initial clinical trials for therapeutics in adult healthy volunteers. Available (Online) at: http://www.fda.gov/downloads/Drugs/.../Guidances/UCM078932.pdf (accessed December 10, 2015).

Food and Drug Administration (FDA). 2010. Information Sheet Guidance for Sponsors, Clinical Investigators and IRBs. Available (Online) at: http://www.fda.gov/downloads/RegulatoryInformation/Guidances/UCM214282.pdf (accessed December 10, 2015).

Food and Drug Administration (FDA). 2014. Investigational New Drug Application. Available (Online) at: http://www.fda.gov/Drugs/DevelopmentApprovalProcess/HowDrugsareDevelopedandApproved/ApprovalApplications/InvestigationalNewDrugINDApplication/default.htm (accessed January 19, 2016).

Food and Drug Administration (FDA); Center for Drug Evaluation and Research. 1995. Guidance for Industry: content and format of Investigational New Drug Applications (INDs) for phase 1 studies of drugs, including well-characterized, therapeutic, biotechnology-derived products. Available (Online) at: http://www.fda.gov/downloads/drugs/guidancecomplianceregulatoryinformation/guidances/ucm071597.pdf (accessed June 14, 2016).

Food and Drug Administration (FDA); Center for Drug Evaluation and Research. 2000. Guidance for Industry: waiver of *in vivo* bioavailability and bioequivalence studies for immediate-release solid oral dosage forms based on a Biopharmaceutics Classification System. Available (Online) at: http://www.fda.gov/OHRMS/DOCKETS/98fr/3657gd3.pdf (accessed November 12, 2015).

Food and Drug Administration (FDA); Center for Drug Evaluation and Research. 2003. Guidance for Industry: bioavailability and bioequivalence studies for orally administered drug products—general considerations. Available (Online) at: http://www.fda.gov/ohrms/dockets/ac/03/briefing/3995B1_07_GFI-BioAvail-BioEquiv.pdf (accessed November 12, 2015).

Food and Drug Administration (FDA); Center for Drug Evaluation and Research. 2008. Guidance for Industry: cGMP for phase 1 investigational drug. Available (Online) at: http://www.fda.gov/downloads/drugs/guidancecomplianceregulatoryinformation/guidances/ucm070273.pdf (accessed May 10, 2016).

Food and Drug Administration (FDA); Center for Drug Evaluation and Research. 2015. Guidance for Industry: waiver of *in vivo* bioavailability and bioequivalence studies for immediate-release solid oral dosage forms based on a Biopharmaceutics Classification System. Available (Online) at: http://www.fda.gov/ForPatients/Approvals/Drugs/ucm405622.htm (accessed January 15, 2016).

Hariharan M, Ganorkar L, Amidon G, Cavallo A, Gatti P, Hageman B, Choo I, Miller J, Shah U. 2003. Reducing the time to develop and manufacture formulations for first oral dose in humans. *Pharm Technol.* 27(10): 68–84.

Hay M, Thomas D, Craighead J, Economides C, Rosenthal J. 2014. Clinical development success rates for investigational drugs. *Nat Biotechnol.* 32: 40–51.

Hunt H, Belanoff, J, Donaldson K, Combs D, Strem M, Zann V, Leung P, Sweet S, Connor A. 2015. Assessment of safety, tolerability, pharmacokinetics, proof of concept and proof of pharmacological effect of orally administered CORT125134 using an adaptive clinical protocol. Quotient Clinical, Presented at the 2015 AAPS Annual Meeting & Exposition, October 25–29, Orange County Convention Center, Orlando, FL

International Conference on Harmonisation (ICH). 1995. Structure and content of Clinical Study Reports E3. Available (Online) at: http://www.ich.org/fileadmin/Public_Web_Site/ICH_Products/Guidelines/Efficacy/E3/E3_Guideline.pdf (accessed January 11, 2016).

International Conference on Harmonisation (ICH). 1996a. Guideline for good clinical practice E6(R1). Available (Online) at: http://www.ich.org/fileadmin/Public_Web_Site/ICH_Products/Guidelines/Efficacy/E6/E6_R1_Guideline.pdf (accessed January 11, 2016).

International Conference on Harmonisation (ICH). 1996b. Stability testing: photostability testing of new drug substances and products Q1B. Available (Online) at: http://www.ich.org/fileadmin/Public_Web_Site/ICH_Products/Guidelines/Quality/Q1B/Step4/Q1B_Guideline.pdf (accessed June 11, 2016).

International Conference on Harmonisation (ICH). 1998. Ethnic factors in the acceptability of foreign clinical data E5(R1). Available (Online) at: http://www. ich.org/fileadmin/Public_Web_Site/ICH_Products/Guidelines/Efficacy/E5_R1/ Step4/E5_R1__Guideline.pdf (accessed February 14, 2016).

International Conference on Harmonisation (ICH). 1999. Specifications: test procedures and acceptance criteria for new drug substances and new drug products: chemical substances Q6A. Available (Online) at: http://www.ich.org/ fileadmin/Public_Web_Site/ICH_Products/Guidelines/Quality/Q6A/Step4/ Q6Astep4.pdf (accessed June 13, 2016).

International Conference on Harmonisation (ICH). 2002a. The common technical document for the registration of pharmaceuticals for human use: Quality—M4Q(R1). Available (Online) at: http://www.ich.org/fileadmin/ Public_Web_Site/ICH_Products/CTD/M4_R1_Quality/M4Q__R1_.pdf (accessed June 13, 2016).

International Conference on Harmonisation (ICH). 2002b. Bracketing and matrixing designs for stability testing of new drug substances and products—Q1D. Available (Online) at: http://www.ich.org/fileadmin/ Public_Web_Site/ICH_Products/Guidelines/Quality/Q1D/Step4/Q1D_ Guideline.pdf (accessed June 13, 2016).

International Conference on Harmonisation (ICH). 2003. Stability testing of new drug substances and products Q1A(R2). Available (Online) at: http://www.ich. org/fileadmin/Public_Web_Site/ICH_Products/Guidelines/Quality/Q1A_R2/ Step4/Q1A_R2__Guideline.pdf (accessed June 5, 2016).

International Conference on Harmonisation (ICH). 2005. Validation of analytical procedures: text and methodology Q2(R1). Available (Online) at: http://www. ich.org/fileadmin/Public_Web_Site/ICH_Products/Guidelines/Quality/Q2_R1/ Step4/Q2_R1__Guideline.pdf (accessed June 13, 2016).

International Conference on Harmonisation (ICH). 2006a. Impurities in new drug substances Q3A(R2). Available (Online) at: http://www.ich.org/fileadmin/ Public_Web_Site/ICH_Products/Guidelines/Quality/Q3A_R2/Step4/Q3A_ R2__Guideline.pdf (accessed June 13, 2016).

International Conference on Harmonisation (ICH). 2006b. Impurities in new drug products Q3B(R2). Available (Online) at: http://www.ich.org/fileadmin/ Public_Web_Site/ICH_Products/Guidelines/Quality/Q3B_R2/Step4/Q3B_ R2__Guideline.pdf (accessed June 13, 2016).

International Conference on Harmonisation (ICH). 2009a. Guidance on nonclinical safety studies for the conduct of human clinical trials and marketing authorization for pharmaceuticals M3(R2). Available (Online) at: http://www. ich.org/fileadmin/Public_Web_Site/ICH_Products/Guidelines/Multidisciplinary/ M3_R2/Step4/M3_R2__Guideline.pdf (accessed January 12, 2016).

International Conference on Harmonisation (ICH). 2009b. Nonclinical evaluation for anticancer pharmaceuticals S9. Available (Online) at: http://www.ich.org/ fileadmin/Public_Web_Site/ICH_Products/Guidelines/Safety/S9/Step4/ S9_Step4_Guideline.pdf (accessed February 13, 2016).

International Conference on Harmonisation (ICH). 2009c. Pharmaceutical development Q8(R2). Available (Online) at: http://www.ich.org/fileadmin/Public_Web_Site/ICH_Products/Guidelines/Quality/Q8_R1/Step4/Q8_R2_Guideline.pdf (accessed June 13, 2016).

Kane Z, Shao J, Christopher R. 2015. Utilisation of RapidFACT strategies to evaluate and develop an in vitro/in vivo correlation (IVIVC) for modified release formulations of lorcaserin HCl. Presented at: *Annual Exposition of the Controlled Release Society (CRS). Edinburgh, UK, July 26–29, 2015.*

Latha R S, Lakshmi P K. 2012. Electronic tongue: an analytical gustatory tool. *J Adv Pharm Technol Res.* 3: 3–8.

McDermott J, Scholes P. 2015. Formulation design space: a proven approach to maximize flexibility and outcomes within early clinical development. *Ther Deliv.* 6(11): 1269–1278.

McDermott J, Connor A, Sidhu S, Mayes B, Moussa A, Ganga S. 2014. Rapid formulation development and clinical evaluation of enabled formulations of IDX-719. Presented at: *American Association of Pharmaceutical Sciences (AAPS) Conference, San Diego, CA, November 2–6, 2014.*

Medicines and Healthcare products Regulatory Agency (MHRA). 2014a. Clinical trials for medicines: apply for authorisation in the UK. Available (Online) at: https://www.gov.uk/guidance/clinical-trials-for-medicines-apply-for-authorisation-in-the-uk (accessed June 13, 2016).

Medicines and Healthcare products Regulatory Agency (MHRA). 2014b. Apply for manufacturer or wholesaler of medicines licenses. Available (Online) at: https://www.gov.uk/guidance/apply-for-manufacturer-or-wholesaler-of-medicines-licences (accessed November 9, 2015).

Medicines and Healthcare products Regulatory Agency (MHRA). 2014c. Good clinical practice for clinical trials. Available (Online) at: https://www.gov.uk/guidance/mhra-phase-i-accreditation-scheme (accessed November 9, 2015).

Millington A D, Chaplan S R, Aguilar Z, Fisher D M, Collier J, Mannens G, Goyvaerts N, Peddareddigari V, Takimoto C, Scholes P. 2015. Rapid evaluation of novel small molecule c-Met tyrosine kinase inhibitor in healthy subjects. Quotient Clinical, Nottingham, UK. Poster CT309 Presented at: *2015 Annual Meeting of the American Association for Cancer Research (AACR), Philadelphia, PA, April 18–22, 2015.*

Muller P, Milton M, Lloyd P, Sims J, Brennan F. 2009. The minimum anticipated biological effect level (MABEL) for selection of first human dose in clinical trials with monoclonal antibodies. *Curr Opin Biotechnol.* 20(6): 722–729.

Murray O, Dang W, Bergstrom D. 2004. Using an electronic tongue to optimize taste-masking in a lyophilized orally disintegrating tablet formulation. *Pharm Technol.* 2004: 42–52. Available (Online) at: http://www.alpha-mos.com/pdf/en/articles/Alpha-MOS_Articles_0408_Pharmtec_2.pdf (accessed November 30, 2015).

Namour F, Vanhoutte F, Beetens J, Blockhuys S, Weer M, Wigerinck P. 2012. Pharmacokinetics, safety and tolerability of GLPG0259 in healthy subjects. *Drugs R&D.* 12(3): 141–163.

O'Quigley J, Pepe M, Fisher L. 1990. Continual reassessment method: a practical design for phase 1 clinical trials in cancer. *Biometrics.* 46(1): 33–48.

Page S, Persch A. 2013. Recruitment, retention and blinding in clinical trials. *Am J Occup Ther.* 67(2): 154–161.

Patrick L J, Connor A L, Mair S J, Collier J E, Rosano M, Hanrahan J. 2012. Design and implementation of a multi-part, flexible protocol to assess the tolerability and pharmacodynamic effects of PUR118 in healthy subjects and COPD patients. Quotient Clinical, Nottingham, UK. *Eur Respir J.* 40: P2155.

Peltonen L, Hirvonen J. 2010. Pharmaceutical nanocrystals by nanomilling: critical process parameters, particle fracturing and stabilization methods. *J Pharm Pharmacol.* 64(11): 1569–1579.

Repka M. 2009. Hot-melt extrusion. *Am Pharm Rev,* October 1, 2009. Available (Online) at: http://www.americanpharmaceuticalreview.com/Featured-Articles/118927-Hot-Melt-Extrusion (accessed January 9, 2016).

Rogatko A, Schoeneck D, Jonas W, Tighiouart M, Khuri F, Porter A. 2007. Translation of innovative designs into phase I trials. *J Clin Oncol.* 25(31): 4982–4986.

Sakpal T. 2010. Sample size estimation in clinical trials. *Perspect Clin Res.* 1(2): 67–69.

Salminen E K, Salminen S J, Kwasowski P, Marks V, Koivistoinen P E. 1989. Xylitol vs glucose: effect on the rate of gastric emptying and motilium, insulin, and gastric inhibitory polypeptide release. *Am J Clin Nutr.* 49(6): 1228–1232.

Shivaani K, Gutierrez M, Doroshow J H, Murgo A J. 2006. Drug development in oncology: classical cytotoxics and molecularly targeted agents. *Br J Clin Pharmacol.* 62(1): 15–26.

Simon R, Freidlin B, Rubinstein L, Arbuck S, Collins J, Christian M. 1997. Accelerated titration designs for phase I clinical trials in oncology. *J Natl Cancer Inst.* 89(15): 1138–1147.

Sistla A, Sunga A, Phung K, Koparkar A, Shenoy N. 2004. Powder-in-bottle formulation of SU011248. Enabling rapid progression into human clinical trials. *Drug Dev Ind Pharm.* 30: 19–25.

Sollohub K, Cal K. 2010. Spray drying technique: II. Current applications in pharmaceutical technology. *J Pharm Sci.* 99(2): 587–597.

Soto J, Sheng Y, Standing J F, Orlu Gul M, Tuleu C. 2015. Development of a model for robust and exploratory analysis of the rodent brief-access taste aversion data. *Eur J Pharm Biopharm.* 91: 47–51.

Tablets & Capsules. 2009. Micro-dosing equipment fills niche in R&D, clinical trial materials. Available (Online) at: http://www.capsugel.com/media/library/micro-dosing-equipment-fills-niche-in-randd-clinical-trial-materials.pdf (accessed December 19, 2015).

Tourneau C, Lee J, Siu, L. 2009. Dose escalation methods in phase I cancer clinical trials. *J Natl Cancer Inst.* 101(10): 708–720.

World Medical Association. 1964. Declaration of Helsinki—ethical principles for research involving human subjects. Available (Online) at: http://www.wma.net/en/30publications/10policies/b3/index.html.pdf?print-media-type&footer-right=[page]/[toPage] (accessed December 22, 2015).

Yin G, Yuan Y. 2009. Bayesian model averaging continual reassessment method in phase I clinical trials. *J Am Stat Assoc.* 104(487): 954–968.

Yu L X, Amidon G L, Polli J E, Zhao H, Mehta M U, Conner D P, Shah V P, Lesko L J, Chen M L, Lee V H, Hussain A S. 2002. Biopharmaceutics classification system: the scientific basis for biowaiver extensions. *Pharm Res.* 19: 921–925.

Index

Oral Formulation Roadmap from Early Drug Discovery to Development,
First Edition. Edited by Elizabeth Kwong.
© 2017 John Wiley & Sons, Inc. Published 2017 by John Wiley & Sons, Inc.